GCSE
IN A
WEEK

Maths

Fiona Mapp

Revision Planner

DAY 1

Page	Time	Topic	Date	Time	Completed
4	15 mins	Prime Factors, HCF and LCM			
6	20 mins	Fractions			
8	20 mins	Percentages			
10	20 mins	Repeated Percentage Change			
12	20 mins	Reverse Percentage Problems			
14	15 mins	Ratio and Proportion			

DAY 2

Page	Time	Topic	Date	Time	Completed
16	15 mins	Rounding and Estimating			
18	15 mins	Indices			
20	20 mins	Standard Index Form			
22	20 mins	Formulae and Expressions 1			
24	20 mins	Formulae and Expressions 2			
26	15 mins	Brackets and Factorisation			
28	15 mins	Equations 1			

DAY 3

Page	Time	Topic	Date	Time	Completed
30	20 mins	Equations 2			
32	20 mins	Simultaneous Linear Equations			
34	15 mins	Sequences			
36	15 mins	Inequalities			
38	20 mins	Straight-line Graphs			
40	25 mins	Curved Graphs			

DAY 4

Page	Time	Topic	Date	Time	Completed
42	15 mins	Distance–Time Graphs			
44	20 mins	Constructions			
46	20 mins	Loci			
48	15 mins	Angles			
50	15 mins	Bearings			
52	20 mins	Translations and Reflections			

Prime Factors, HCF and LCM

Prime Factors

Apart from **prime numbers**, any whole number greater than 1 can be written as a product of **prime factors**. This means the number is written using only prime numbers multiplied together.

A prime number has only two factors, 1 and itself. 1 is not a prime number.

The prime numbers up to 20 are:

2, 3, 5, 7, 11, 13, 17, 19

The diagram below shows the prime factors of 60.

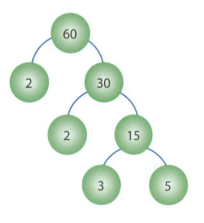

🔵 Divide 60 by its first prime factor, 2.

🔵 Divide 30 by its first prime factor, 2.

🔵 Divide 15 by its first prime factor, 3.

🔵 We can now stop because the number 5 is prime.

As a product of its prime factors, 60 may be written as:

$$60 = 2 \times 2 \times 3 \times 5$$

or

$$60 = 2^2 \times 3 \times 5$$

Highest Common Factor (HCF)

The highest factor that two numbers have in common is called the **HCF**.

Example

Find the HCF of 60 and 96.

🔵 Write the numbers as products of their prime factors.

$$60 = 2 \times 2 \qquad \times 3 \times 5$$

$$96 = 2 \times 2 \times 2 \times 2 \times 2 \times 3$$

🔵 Ring the factors that are common.

$$60 = 2 \times 2 \qquad \times 3 \times 5$$

$$96 = 2 \times 2 \times 2 \times 2 \times 2 \times 3$$

🔵 These give the HCF $= 2 \times 2 \times 3$

$$= \mathbf{12}$$

Lowest (Least) Common Multiple (LCM)

The **LCM** is the lowest number that is a multiple of two numbers.

Example

Find the LCM of 60 and 96.

- Write the numbers as products of their prime factors.

  $60 = 2 \times 2 \qquad\qquad \times 3 \times 5$

  $96 = 2 \times 2 \times 2 \times 2 \times 2 \times 3$

- 60 and 96 have a common factor of $2 \times 2 \times 3$, so it is only counted once.

  $60 = \boxed{2} \times \boxed{2} \qquad\qquad \times \boxed{3} \times 5$

  $96 = \boxed{2} \times \boxed{2} \times 2 \times 2 \times 2 \times \boxed{3}$

- The LCM of 60 and 96 is

  $2 \times 2 \times 2 \times 2 \times 2 \times 3 \times 5$

  $= \mathbf{480}$

SUMMARY

- Any whole number greater that 1 can be written as a product of its prime factors, apart from prime numbers themselves (1 is not prime).

- The highest factor that two numbers have in common is called the Highest Common Factor (HCF).

- The lowest number that is a multiple of two numbers is called the Lowest (Least) Common Multiple (LCM).

QUESTIONS

QUICK TEST

1. Write these numbers as products of their prime factors:

   a. 50

   b. 360

   c. 16

2. Decide whether these statements are true or false.

   a. The HCF of 20 and 40 is 4.

   b. The LCM of 6 and 8 is 24.

   c. The HCF of 84 and 360 is 12.

   d. The LCM of 24 and 60 is 180.

EXAM PRACTICE

1. Find the Highest Common Factor of 120 and 42.

2. Buses to St Albans leave the bus station every 20 minutes. Buses to Hatfield leave the bus station every 14 minutes.

   A bus to St Albans and a bus to Hatfield both leave the bus station at 10 am. When will buses to both St Albans and Hatfield next leave the bus station at the same time?

Fractions

A **fraction** is part of a whole number. The top number is the **numerator** and the bottom number is the **denominator**.

There are four rules of fractions:

Addition $+$

You need to change the fractions so that they have the same denominator.

Example

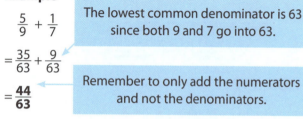

$$\frac{5}{9} + \frac{1}{7}$$

The lowest common denominator is 63 since both 9 and 7 go into 63.

$$= \frac{35}{63} + \frac{9}{63}$$

$$= \frac{44}{63}$$

Remember to only add the numerators and not the denominators.

Subtraction $-$

You need to change the fractions so that they have the same denominator.

Example

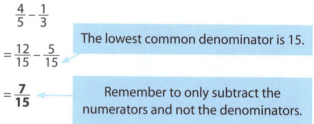

$$\frac{4}{5} - \frac{1}{3}$$

The lowest common denominator is 15.

$$= \frac{12}{15} - \frac{5}{15}$$

$$= \frac{7}{15}$$

Remember to only subtract the numerators and not the denominators.

FOUR RULES OF FRACTIONS

Multiplication $\times$

Before starting, write out whole or mixed numbers as improper fractions (also known as top-heavy fractions).

Example

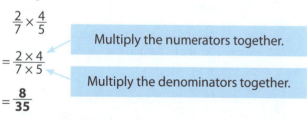

$$\frac{2}{7} \times \frac{4}{5}$$

Multiply the numerators together.

$$= \frac{2 \times 4}{7 \times 5}$$

Multiply the denominators together.

$$= \frac{8}{35}$$

Division $\div$

Before starting, write out whole or mixed numbers as improper fractions (also known as top-heavy fractions).

Example

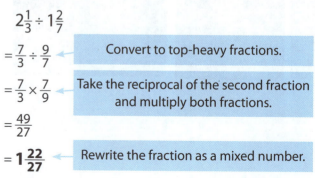

$$2\frac{1}{3} \div 1\frac{2}{7}$$

Convert to top-heavy fractions.

$$= \frac{7}{3} \div \frac{9}{7}$$

Take the reciprocal of the second fraction and multiply both fractions.

$$= \frac{7}{3} \times \frac{7}{9}$$

$$= \frac{49}{27}$$

$$= 1\frac{22}{27}$$

Rewrite the fraction as a mixed number.

Fraction Problems

You may need to solve problems involving fractions.

Examples

1. A school has 1400 pupils. 740 pupils are boys.

   $\frac{3}{5}$ of the boys study French,

   $\frac{1}{4}$ of the girls study French.

   Work out the total number of pupils in the school who study French.

   $\frac{3}{5} \times 740 = 444$ boys study French
   > Work out the number of boys who study French.

   $1400 - 740 = 660$ are girls
   > Work out the number of girls in the school.

   $\frac{1}{4} \times 660 = 165$ girls study French

   $444 + 165 = \textbf{609 pupils study French}$

2. Charlotte's take-home pay is £930. She gives her mother $\frac{1}{3}$ of this and spends $\frac{1}{5}$ of the £930 on going out. What fraction of the £930 is left?

   Give your answer as a fraction in its simplest form.

   $\frac{1}{3} + \frac{1}{5}$
   > This is a simple addition of fractions question.

   $= \frac{5}{15} + \frac{3}{15}$
   > Write the fractions with a common denominator.

   $= \frac{8}{15}$

   $1 - \frac{8}{15}$
   > The question asks for the fraction of the money that is left, so subtract $\frac{8}{15}$ from 1.

   $= \frac{7}{15}$
   > The fraction is in its simplest form.

SUMMARY

- To add or subtract fractions, write them using the same denominator.
- To multiply fractions, multiply the numerators and multiply the denominators.
- To divide fractions, take the reciprocal of the second fraction and multiply the fractions together.
- When multiplying and dividing fractions, write out whole or mixed numbers as top-heavy fractions before you begin the calculation.

QUESTIONS

QUICK TEST

1. Work out the following:

   a. $\frac{2}{3} + \frac{1}{5}$

   b. $2\frac{6}{7} - \frac{1}{3}$

   c. $\frac{2}{9} \times \frac{5}{7}$

   d. $\frac{3}{11} \div \frac{22}{27}$

2. Work out the following:

   a. $2\frac{1}{2} + 3\frac{1}{5}$

   b. $2\frac{7}{10} - 1\frac{1}{9}$

   c. $3\frac{1}{5} \times \frac{2}{15}$

   d. $5\frac{1}{4} \div \frac{3}{8}$

EXAM PRACTICE

1. In a magazine $\frac{3}{7}$ of the pages have advertisements on them. Given that 12 pages have advertisements on them, work out the number of pages in the magazine.

2. Rosie watches two television programmes. The first programme is $\frac{3}{4}$ of an hour and the second is $2\frac{2}{3}$ hours long. Work out the total length of the two programmes.

Percentages

Percentages are fractions with a denominator of 100.

This is the percentage sign:

Percentage of a Quantity

OF means multiply. To find the percentage of a quantity, write the percentage as a fraction with a denominator of 100 and multiply by the quantity.

Example
Find 30% of 80 kg.

$\frac{30}{100} \times 80 = 24\,kg$

On the calculator, key in:

| 30 | ÷ | 100 | × | 80 | = |

$30\% = \frac{30}{100} = 0.3$, which is known as the **multiplier**.

For the **non-calculator** paper:

● find 10% by dividing by 10

10% of 80 kg

$= \frac{80}{10}$

$= 8\,kg$

● then multiply by 3 to get 30%

3×8

$= \mathbf{24\,kg}$

Increasing and Decreasing Quantities by a Percentage

Percentages appear in everyday life and you will often need to find the value of quantities after a percentage increase or decrease.

Examples

1. An MP3 player costs £165. In a sale it is reduced by 15%. Work out the cost of the MP3 player in the sale.

   Method 1

   15% of £165

   $= \frac{15}{100} \times 165$

   $= £24.75$

   Price of the MP3 player in the sale

   $= £165 - £24.75 = \mathbf{£140.25}$

   Method 2 (using a multiplier)

   $1 - 0.15 = 0.85$

   0.85×165

   | 0.85 is the multiplier since the price is going down. |

   $= \mathbf{£140.25}$

2. Louisa works out the cost of her gas bill. At the start of a three-month period the gas meter reading was 12 447 units. At the end of the three-month period the gas meter reading was 12 721.

   Each unit of gas used costs 47p. VAT is charged at 5%. Work out the total cost of Louisa's gas bill.

   Units used   12 721 − 12 447
               = 274

   Cost of gas   274 × 47p
               = £128.78

   VAT at 5%
   Multiplier $1 + 0.05 = 1.05$
               1.05×128.78

   | 1.05 is the multiplier since the price is going up. |

   Gas bill $= \mathbf{£135.22}$

   | Round to the nearest penny. |

One Quantity as a Percentage of Another

To express one quantity as a percentage of another, divide the first quantity by the second quantity and multiply by 100%.

Example

Matthew got 46 out of 75 in a Science test. He got 65% in a Maths test. In which test did he do the best?

Work out the percentage he got in the Science test.

$\frac{46}{75} \times 100\%$ | Make a fraction and multiply by 100%.

$= 61.\dot{3}\%$

On the calculator, key in:

[46] [÷] [75] [×] [100] [=]

In the Maths test Matthew got 65%, so he did better in the Maths test than the Science test.

SUMMARY

- Percentages are fractions with a denominator of 100.

- To find the percentage of a quantity, write the percentage as a fraction with a denominator of 100 and multiply by the quantity.

- To express one quantity as a percentage of another, divide the first quantity by the second quantity and multiply by 100%.

- OF means multiply.

QUESTIONS

QUICK TEST

1. Without using a calculator, work out the following:

   a. 12% of 50 kg
   b. 30% of £2000
   c. 5% of £60
   d. 35% of 720 g

2. Without using a calculator, change each fraction into a percentage:

   a. $\frac{16}{50}$
   b. $\frac{46}{200}$
   c. $\frac{15}{20}$
   d. $\frac{21}{25}$

3. Reduce £225 by 20%.

4. Express 32 as a percentage of 40.

EXAM PRACTICE

1. Jonathan is buying a new television. He sees three different advertisements for the same television.

BEST TV SHOP	DRYMONS	MARK'S ELECTRICALS
Normal Price £556 SALE: $\frac{1}{5}$ off Normal Price	TV Normal price £495 Sale 10% off	TV £385 PLUS VAT at 20%

   Jonathan wants to buy the cheapest television. From which shop should Jonathan purchase his television? You must show full working out and give a reason for your answer.

2. Madeleine earns £48 500 per year. She does not pay income tax on the first £7475 of her salary (which is deducted before tax is calculated). She then pays the basic rate of tax at 20% on her earnings up to £35 000, and the higher rate of tax at 40% on anything over £35 000. How much tax will Madeleine pay?

Repeated Percentage Change

Percentage Change

> Percentage change = $\frac{\text{change}}{\text{original}} \times 100\%$

Examples

1. Tammy bought a flat for £185 000. Three years later she sold it for £242 000. What is her percentage profit?

   Profit is £242 000 – £185 000

   $= £57 000$

   Percentage profit is $\frac{57\,000}{185\,000} \times 100\%$

   $= \mathbf{30.8\%}$ (3sf)

2. Jackie bought a car for £12 500 and sold it two years later for £7250. Work out her percentage loss.

   Loss is £12 500 – £7250

   $= £5250$

   Percentage loss is $\frac{5250}{12\,500} \times 100\%$

   $= \mathbf{42\%}$

Repeated Percentage Change

A quantity can increase or decrease in value each year by a different percentage. These quantities will change in value at the end of each year. To calculate repeated percentage change, two methods are explained in the example below.

Example

A car was bought for £12 500. Each year it depreciated in value by 15%. What was the car worth after three years?

> You must remember **not** to do 3 × 15% = 45% reduction over 3 years!

Method 1

● Find 100% – 15% = 85% of the value of the car first.

   Year 1: $\frac{85}{100} \times £12\,500 = £10\,625$

● Then work out the value year by year. (£10 625 depreciates in value by 15%.)

   Year 2: $\frac{85}{100} \times £10\,625 = £9031.25$

   (£9301.25 depreciates in value by 15%.)

   Year 3: $\frac{85}{100} \times £9031.25 = \mathbf{£7676.56}$ (2dp)

Method 2

● A quick way to work this out is by using a multiplier.

● Finding 85% of the value of the car is the same as multiplying by 0.85

   Year 1: 0.85 × £12 500 = £10 625

   Year 2: 0.85 × £10 625 = £9031.25

   Year 3: 0.85 × £9031.25 = **£7676.56** (2dp)

● This is the same as working out $(0.85)^3 \times £12\,500 = \mathbf{£7676.56}$ (2dp)

Compound Interest

Compound interest is where the bank pays interest on the interest already earned as well as on the original money.

Example
Becky has £3200 in her savings account and compound interest is paid at 3.2% per annum. How much will she have in her account after four years?

100% + 3.2% = 103.2%

= 1.032   This is the multiplier.

Year 1: 1.032 × £3200 = £3302.40

Year 2: 1.032 × £3302.40 = £3408.08

Year 3: 1.032 × £3408.08 = £3517.14

Year 4: 1.032 × £3517.14 = £3629.68

Total = **£3629.68** (2dp)

A quicker way is to multiply £3200 by $(1.032)^4$

Number of years

£3200 × $(1.032)^4$ = **£3629.68** (2dp)

Original   Multiplier

Simple Interest

Simple interest is the interest paid each year. It is the same amount each year.

The simple interest on £3200 invested for four years at 3.2% per annum would be:

$\frac{3.2}{100}$ × 3200 = £102.40 for one year

Interest over four years would be 4 × £102.40 = £409.60

Total in account after 4 years would be £3609.60

SUMMARY

- **A quick way to work out repeated percentage change is to use a multiplier.**

- **Compound interest is where the bank pays interest on the interest already earned as well as on the original money.**

- **Simple interest is the interest paid each year. It is the same amount each year.**

QUESTIONS

QUICK TEST

1.  A car is bought for £8500. Two years later it is sold for £4105. Work out the percentage loss. Give your answer to 3 significant figures.

2. A flat was bought for £85 000 in 2005. The flat rose in value by 12% in 2006 and 28% in 2007. How much was the flat worth at the end of 2007?

EXAM PRACTICE

1. Shamil invests £3000 in each of two bank accounts. The terms of the bank accounts are shown below.

Savvy Saver	Money Grows
Simple interest at 2.5% per annum.	Compound interest at 2.5% per annum.

Shamil says that he will earn the same amount of interest from both bank accounts in two years.

Decide whether Shamil is correct. You must show full working to justify your answer.

Reverse Percentage Problems

In reverse percentage problems you are given the final amount after a percentage increase or decrease. You have to then find the value of the original quantity. These are quite tricky, so think carefully.

Example 1

The price of a television is reduced by 15% in the sales. It now costs £352.75. What was the original price?

15% off

● The sale price is 100% − 15% = 85% of the pre-sale price (x)

● 85% = 0.85   This is the multiplier.

● $0.85 \times x = £352.75$

$$x = \frac{£352.75}{0.85}$$

Original price is **£415**

Check:

original price — × 0.85 → new price

original price ← ÷ 0.85 — new price

Does the answer sound sensible?
Is the original price more than the sale price?

Example 2

A telephone bill costs £169.20 including VAT at 20%. What is the cost of the bill without the VAT?

● The telephone bill of £169.20 represents 100% + 20% = 120% of the original bill (x).

● 120% = 1.20   This is the multiplier.

● $1.2 \times x = £169.20$

$$x = \frac{£169.20}{1.2}$$

Original bill is **£141**

Check:

original bill — × 1.2 → new bill

original bill ← ÷ 1.2 — new bill

Example 3

The price of a washing machine is reduced by 5% in the sales. It now costs £323. What was the original price?

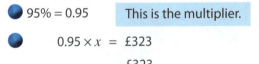

- The sale price is 100% – 5% = 95% of the pre-sale price (x).

- 95% = 0.95   This is the multiplier.

- $$0.95 \times x = £323$$

  $$x = \frac{£323}{0.95}$$

Original price is **£340**

Check:

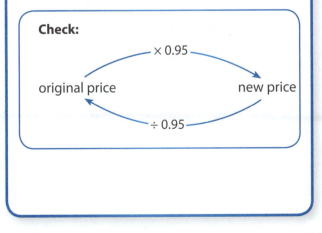

× 0.95

original price   new price

÷ 0.95

QUESTIONS

QUICK TEST

1. 🔢 Each item listed below includes VAT at 20%. Work out the original price of the item.

   a. A pair of shoes: £69

   b. A coat: £152.40

   c. A suit: £285

   d. A television: £525

EXAM PRACTICE

1. 🔢 In a sale, normal prices are reduced by 12%. The sale price of a television is £220.

   Work out the normal price of the television.

2. 🔢 Joseph says that the original price of a CD player, which now costs £60 after a 15% reduction, was £70.59

   Is Joseph correct? Show your working.

Ratio and Proportion

Sharing a Quantity in a Given Ratio

A **ratio** is used to compare two or more related quantities.

To share an amount in a given ratio, add up the individual parts and then divide the amount by this number to find one part.

Example

£155 is divided in the ratio of 2 : 3 between Daisy and Tom. How much does each receive?

2 + 3 = 5 parts   Add up the total parts.

5 parts = £155

1 part = £155 ÷ 5   Work out what one part is worth.

= £31

So Daisy gets 2 × £31 = **£62**
and Tom gets 3 × £31 = **£93**

Check: £62 + £93
= £155 ✔

Exchange Rates

Two quantities are in **direct proportion** when both quantities increase at the same rate.

Example

Samuel went on holiday to Spain. He changed £350 into Euros. The exchange rate was £1 = €1.16. How many Euros did Samuel receive?

£1 = €1.16 so

£350 = 350 × 1.16

= **€406**

Best Buys

Use unit amounts to help you decide which is the better value for money.

Example

The same brand of breakfast cereal is sold in two different-sized packets. Which packet represents better value for money?

£3.15

£1.65

500g

125g

Find the cost per gram for both boxes of cereal.

125 g costs £1.65 so $\frac{165}{125}$ = 1.32p per gram

500 g costs £3.15 so $\frac{315}{500}$ = 0.63p per gram

Since the larger box costs less per gram, it represents better value for money.

Increasing and Decreasing in a Given Ratio

When increasing or decreasing in a given ratio, it is sometimes easier to find a unit amount.

Example

A photograph of length 12 cm is to be enlarged in the ratio 4 : 5

What is the length of the enlarged photograph?

$\frac{12}{4} = 3$ cm | Divide 12 by 4 to get 1 part.

$3 \times 5 = \textbf{15 cm}$ | Multiply this by 5 to get the length of the enlarged photograph.

When two quantities are in **inverse proportion**, one quantity increases at the same rate as the other quantity decreases. For example, the time it takes to build a wall increases as the number of builders decreases.

A wall took 4 builders 6 days to build.

Time for 4 builders = 6 days

Time for 1 builder   = 6 × 4 = 24 days

It takes 1 builder four times as long to build the wall.

At the same rate it would take 6 builders $\frac{24}{6} = 4$ days

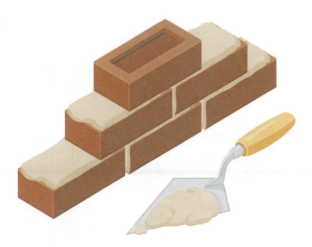

SUMMARY

- A ratio is used to compare two or more related quantities.

- Two quantities are in direct proportion when both quantities increase at the same rate.

- Two quantities are in inverse proportion when one quantity increases at the same rate as the other quantity decreases.

QUESTIONS

QUICK TEST

1. Divide £160 in the ratio 1 : 2 : 5

2. The cost of four ringbinders is £6.72 Work out the cost of 21 ringbinders.

3. A patio took 6 builders 4 days to lay. At the same rate how long would it take 8 builders?

EXAM PRACTICE

1. Toothpaste is sold in three different-sized tubes.

   50 ml is £1.24

   75 ml is £1.96

   100 ml is £2.42

   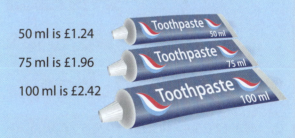

   Which of the tubes of toothpaste is the better value for money? You must show full working in order to justify your answer.

2. Jessica buys a pair of jeans in England for £52. She then goes on holiday to America and sees an identical pair of jeans for $63. The exchange rate is £1 = $1.49. In which country are the jeans cheaper, and by how much?

Rounding and Estimating

Decimal Places

When rounding numbers to a given number of **decimal places** (dp), count the number of places to the right of the decimal point, then look at the next digit on.

If the number is 5 or bigger, round up.
If the number is 4 or smaller, the digit stays the same.

Significant Figures

The first **significant figure** (sf) is the first digit that is not a zero. The 2nd, 3rd… significant figures follow on after the first significant figure. They may or may not be zeros.

The same rules apply as in decimal places.

2.3725 = 2.373 (3dp)

The digit is 5 so round up the 2.

The 2 rounds up to a 3.

6347 = 6350 (3sf)

The digit is bigger than 5, so the 4 rounds to a 5.

You must fill in the end zero(s). This is often forgotten.

Examples
Round the following to the number of decimal places specified in the brackets.

1. 4.6931 (2dp) = **4.69**

   The 3 is less than 5, so the 9 does not change.

2. 27.325 (2dp) = **27.33**

   The 5 has the effect of rounding the 2 to a 3.

3. 149.3867 (3dp) = **149.387**

   The 7 has the effect of rounding the 6 to a 7.

4. 271.74 (1dp) = **271.7**

   The 4 is less than 5, so has no effect on the 7.

Examples
1. Round 9.3156 to…

   a. 3sf = **9.32**

   The 5 has the effect of rounding the 1 to a 2.

   b. 2sf = **9.3**

   The 1 is less than 5, so the 3 does not change.

   c. 1sf = **9**

   The 3 is less than 5, so to 1sf it is 9, since 9.3156 is nearer to 9 than 10.

2. Round 0.735 to…

   a. 2sf = **0.74**

   The 5 has the effect of rounding the 3 to a 4.

   b. 1sf = **0.7**

   The 3 is less than 5, so the 7 does not change.

Estimating

When estimating the answer to a calculation, you must round the number to 1 significant figure.

$$\frac{273 \times 49}{28} \approx \frac{300 \times 50}{30} = 500$$

$\approx$ means approximately equal to

If a measurement is accurate to some given amount, then the true value lies within half a unit of that amount.

If the weight (w) of a cat is 8.3 kg to the nearest tenth of a kilogram, then the weight would lie between 8.25 kg and 8.35 kg.

$$8.25 \leqslant w < 8.35$$

Lower bound   Upper bound

SUMMARY

- To round or correct to a given number of decimal places (dp), count that number of decimal places to the right of the decimal point. Look at the next digit on. If it is 5 or more you need to round up. Otherwise the digit stays the same.

- When measurements are given to a certain degree of accuracy:
  - highest value = upper bound
  - lowest value = lower bound

- For any number, the first significant figure is any number that is not a zero. The 2nd, 3rd... significant figures follow on after the first significant figure. They may or may not be zeros.

QUESTIONS

QUICK TEST

1. Put a ring around the correct answer.
   3724 rounded to 2 significant figures is:

   a. 3800   **b.** 37   **c.** 38   **d.** 3700

2. Decide whether the following statements are true or false.

   a. 4625 rounded to 3sf is 4630

   b. 2.795 rounded to 1dp is 2.7

   c. 0.00527 rounded to 2sf is 0.0053

   d. 37 062 has 4 significant figures

EXAM PRACTICE

1. Work out an estimate for:
   $$\frac{306 \times 2.93}{0.051}$$

2. The weight of a book is 28 grams to the nearest gram. Write down the lower bound of the weight of the book.

Indices

An **index** is sometimes called a **power**.

| The base | $\longrightarrow$ | a^b | $\longleftarrow$ | The index or power |

Laws of Indices

The laws of indices can be used for numbers or in algebra. The base has to be the same when the laws of indices are applied.

$$a^n \times a^m = a^{n+m}$$

$$a^n \div a^m = a^{n-m}$$

$$(a^n)^m = a^{n \times m}$$

$$a^0 = 1$$

$$a^1 = a$$

$$a^{-n} = \frac{1}{a^n}$$

$$a^{\frac{1}{m}} = \sqrt[m]{a}$$

$$a^{\frac{n}{m}} = (\sqrt[m]{a})^n$$

Examples with Numbers

1. Simplify the following, leaving your answers in index notation.

   a. $5^2 \times 5^3 = 5^{2+3} = \mathbf{5^5}$

   b. $8^{-5} \times 8^{12} = 8^{-5+12} = \mathbf{8^7}$

   c. $(2^3)^4 = 2^{3 \times 4} = \mathbf{2^{12}}$

2. Evaluate: Evaluate means to work out.

   a. $4^2 = 4 \times 4 = \mathbf{16}$

   b. $5^0 = \mathbf{1}$

   c. $3^{-2} = \frac{1}{3^2} = \frac{\mathbf{1}}{\mathbf{9}}$

   d. $36^{\frac{1}{2}} = \sqrt{36} = \mathbf{6}$

   e. $8^{\frac{2}{3}} = (\sqrt[3]{8})^2 = 2^2 = \mathbf{4}$

3. Simplify the following, leaving your answers in index form.

   a. $7^2 \times 7^5 = \mathbf{7^7}$

   b. $6^9 \div 6^2 = \mathbf{6^7}$

   c. $\frac{3^7 \times 3^2}{3^{10}} = \frac{3^9}{3^{10}} = \mathbf{3^{-1}}$

   d. $7^9 \div 7^{-10} = \mathbf{7^{19}}$

4. Evaluate:

   a. $3^3 = 3 \times 3 \times 3 = \mathbf{27}$

   b. $7^0 = \mathbf{1}$

   c. $64^{\frac{1}{3}} = \sqrt[3]{64} = \mathbf{4}$

   d. $81^{\frac{1}{2}} = \sqrt{81} = \mathbf{9}$

   e. $5^{-2} = \frac{1}{5^2} = \frac{\mathbf{1}}{\mathbf{25}}$

   f. $\left(\frac{4}{9}\right)^{-2} = \left(\frac{9}{4}\right)^2 = \frac{81}{16} = \mathbf{5\frac{1}{16}}$

Examples with Algebra

1. Simplify the following:

a. $a^4 \times a^{-6} = a^{4-6} = a^{-2} = \dfrac{1}{a^2}$

b. $5y^2 \times 3y^6 = \textbf{15}\textbf{\textit{y}}^{\textbf{8}}$

> The numbers are multiplied.

> The indices are added.

c. $(4x^3)^2 = \textbf{16}\textbf{\textit{x}}^{\textbf{6}}$

> Remember to square the 4 as well.

If in doubt, write it out: $(4x^3)^2 = 4x^3 \times 4x^3$

$$= \textbf{16}\textbf{\textit{x}}^{\textbf{6}}$$

d. $(3x^4y^2)^3 = \textbf{27}\textbf{\textit{x}}^{\textbf{12}}\textbf{\textit{y}}^{\textbf{6}}$

or $3x^4y^2 \times 3x^4y^2 \times 3x^4y^2 = \textbf{27}\textbf{\textit{x}}^{\textbf{12}}\textbf{\textit{y}}^{\textbf{6}}$

e. $(2x)^{-3} = \dfrac{1}{(2x)^3} = \dfrac{\textbf{1}}{\textbf{8}\textbf{\textit{x}}^{\textbf{3}}}$

2. Simplify:

a. $\dfrac{15b^4 \times 3b^7}{5b^2} = \dfrac{45b^{11}}{5b^2} = \textbf{9}\textbf{\textit{b}}^{\textbf{9}}$

b. $\dfrac{16a^2b^4}{4ab^3} = \textbf{4}\textbf{\textit{ab}}$

3. Simplify:

a. $7a^2 \times 3a^2b = \textbf{21}\textbf{\textit{a}}^{\textbf{4}}\textbf{\textit{b}}$

b. $\dfrac{14a^2b^4}{7ab} = \textbf{2}\textbf{\textit{ab}}^{\textbf{3}}$

c. $\dfrac{9x^2y \times 2xy^3}{6xy} = \dfrac{18x^3y^4}{6xy}$

$$= \textbf{3}\textbf{\textit{x}}^{\textbf{2}}\textbf{\textit{y}}^{\textbf{3}}$$

SUMMARY

- Make sure you know and can use all the laws of indices.
- A negative power is the reciprocal of the positive power.
- Fractional indices mean roots.

QUESTIONS

QUICK TEST

1. Simplify the following, leaving your answers in index form.

  a. $6^3 \times 6^5$

  b. $12^{10} \div 12^{-3}$

  c. $(5^2)^3$

  d. $64^{\frac{2}{3}}$

2. Simplify the following:

  a. $2b^4 \times 3b^6$

  b. $8b^{-12} \div 4b^4$

  c. $(3b^4)^2$

  d. $(5x^2y^3)^{-2}$

EXAM PRACTICE

1. Evaluate:

  a. 5^0 **b.** 7^{-2}

  c. $64^{\frac{1}{3}}$ **d.** $27^{-\frac{2}{3}}$

2. Simplify:

  a. $\dfrac{x^4 \times x^7}{x^{15}}$

  b. $\dfrac{3x^4 \times 4x^2}{2x^3}$

Standard Index Form

Standard index form (standard form) is useful for writing very large or very small numbers in a simpler way.

When written in standard form a number will be written as:

A number between 1 and 10 $1 \leqslant a < 10$	$a \times 10^n$ The value of n is the number of places the digits have to be moved to return the number to its original value.

If the number is 10 or more, n is positive.

If the number is less than 1, n is negative.

If the number is between 1 and 10, n is zero.

Examples

1. Write 2 730 000 in standard form.

- 2.73 is the number between 1 and 10 $(1 \leqslant 2.73 < 10)$

- Count how many spaces the digits have to move to restore the original number.
  The digits have moved 6 places to the left because it has been multiplied by 10^6
  2 . 7 3
  2 7 3 0 0 0 0
  So, 2 730 000 = **2.73 × 10⁶**

2. Write 0.000046 in standard form.

- Put the decimal point between the 4 and 6, so the number lies between 1 and 10.

- Move the digits five places to the right to restore the original number.

- The value of n is negative.

  So, 0.000046 = **4.6 × 10⁻⁵**

On a Calculator

 To put a number written in standard form into your calculator you use the following key:

$$\boxed{\text{EXP}} \quad \boxed{\text{EE}} \quad \text{or} \quad \boxed{\times 10^x}$$

For example, $(2 \times 10^3) \times (6 \times 10^7) = 1.2 \times 10^{11}$ would be keyed in as:

$$\boxed{2}\ \boxed{\text{EXP}}\ \boxed{3}\ \boxed{\times}\ \boxed{6}\ \boxed{\text{EXP}}\ \boxed{7}\ \boxed{=}$$

or

$$\boxed{2}\ \boxed{\times 10^x}\ \boxed{3}\ \boxed{\times}\ \boxed{6}\ \boxed{\times 10^x}\ \boxed{7}\ \boxed{=}$$

Doing Calculations

Examples

 Work out the following using a calculator. Check that you get the answers given here.

1. $(6.7 \times 10^7)^3 = \mathbf{3.0 \times 10^{23}}$ (2sf)

2. $\dfrac{(4 \times 10^9)}{(3 \times 10^4)^2} = \mathbf{4.\dot{4}}$

3. $\dfrac{(5.2 \times 10^6) \times (3 \times 10^7)}{(4.2 \times 10^5)^2} = \mathbf{884.4}$ (1dp)

Examples

On a non-calculator paper you can use indices to help work out your answers.

1. $(2 \times 10^3) \times (6 \times 10^7)$

   $= (2 \times 6) \times (10^3 \times 10^7)$

   $= 12 \times 10^{3+7}$

   $= 12 \times 10^{10}$

   $= 1.2 \times 10^1 \times 10^{10}$

   $= \mathbf{1.2 \times 10^{11}}$

2. $(6 \times 10^4) \div (3 \times 10^{-2})$

   $= (6 \div 3) \times (10^4 \div 10^{-2})$

   $= 2 \times 10^{4-(-2)}$

   $= \mathbf{2 \times 10^6}$

3. $(3 \times 10^4)^2$

   $= (3 \times 10^4) \times (3 \times 10^4)$

   $= (3 \times 3) \times (10^4 \times 10^4)$

   $= \mathbf{9 \times 10^8}$

You also need to be able to work out more complex calculations.

Example

The mass of Saturn is 5.7×10^{26} tonnes. The mass of the Earth is 6.1×10^{21} tonnes. How many times heavier is Saturn than the Earth? Give your answer in standard form, correct to 2 significant figures.

$$\frac{5.7 \times 10^{26}}{6.1 \times 10^{21}} = 93\,442.6$$

Now rewrite your answer in standard form.

Saturn is 9.3×10^4 times heavier than the Earth.

SUMMARY

- Numbers in standard form will be written as $a \times 10^n$.
- $1 \leqslant a < 10$
- n is positive when the original number is 10 or more.
- n is negative when the original number is less than 1.
- n is zero when the original number is between 1 and 10.

QUESTIONS

QUICK TEST

1. Write in standard form:

   a. 64 000

   b. 0.00046

2. Without a calculator, work out the following. Leave in standard form.

   a. $(3 \times 10^4) \times (4 \times 10^6)$

   b. $(6 \times 10^{-5}) \div (3 \times 10^{-4})$

3. Work these out on a calculator:

   a. $(4.6 \times 10^{12}) \div (3.2 \times 10^{-6})$

   b. $(7.4 \times 10^9)^2 + (4.1 \times 10^{11})$

EXAM PRACTICE

1. a. Write 40 000 000 in standard form.

   b. Write 6×10^{-5} as an ordinary number.

2. The mass of an atom is 2×10^{-23} grams. What is the total mass of 7×10^{16} of these atoms?

   Give your answer in standard form.

Formulae and Expressions 1

A **term** is a collection of numbers, letters and brackets, all multiplied together, e.g. $6a$, $2ab$, $3(x-1)$.

Expressions are made up of a number of terms, e.g. $a + 6$.

Terms are separated by + and – signs. Each term has a + or – attached to the front of it.

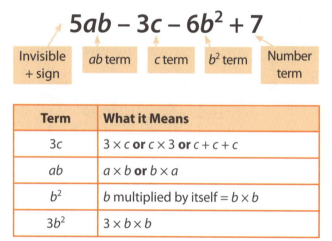

$$5ab - 3c - 6b^2 + 7$$

Invisible + sign	ab term	c term	b^2 term	Number term

Term	What it Means
$3c$	$3 \times c$ **or** $c \times 3$ **or** $c + c + c$
ab	$a \times b$ **or** $b \times a$
b^2	b multiplied by itself $= b \times b$
$3b^2$	$3 \times b \times b$

🔴 $a \div 2$ can be written as $\dfrac{a}{2}$

🔴 $c \times a \times 5 = 5ac$; the number usually comes first and then the letters in alphabetical order

🔴 $3a^2$ is not the same as $(3a)^2$

$3a^2$ is 3 lots of just a^2

$(3a)^2$ is 3 multiplied by a, then all of it squared

Collecting Like Terms

Expressions can be simplified by **collecting like terms**.

Simplify means make the expression simpler.

You can only collect together terms that include exactly the same letter combinations.

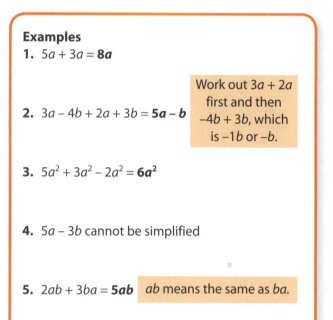

Examples
1. $5a + 3a = \mathbf{8a}$

   > Work out $3a + 2a$ first and then $-4b + 3b$, which is $-1b$ or $-b$.

2. $3a - 4b + 2a + 3b = \mathbf{5a - b}$

3. $5a^2 + 3a^2 - 2a^2 = \mathbf{6a^2}$

4. $5a - 3b$ cannot be simplified

5. $2ab + 3ba = \mathbf{5ab}$ ab means the same as ba.

$$a^2 - b^2 = (a-b)(a+b$$

Writing Formulae

$b = a + 6$ is called a **formula**. The value of b depends on the value of a.

Sometimes you will need to write your own formulae.

Example

My brother is 3 years older than me. My mother is 3 times as old as me.

If I am n years old, write expressions for my brother's and mother's ages.

If I am 15, my brother is $15 + 3 = 18$

> Try using numbers first.

So if I am n years old my brother is $n + 3$ years old.

If I am 15, my mother is $3 \times 15 = 45$

So if I am n years old my mother is $3 \times n$ or $3n$ years old.

Write down a formula for the sum (s) of the ages of me, my mother and brother.

$s = n + n + 3 + 3n$

$s = \mathbf{5n + 3}$

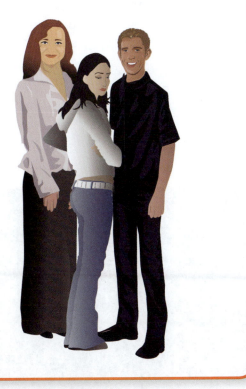

QUESTIONS

QUICK TEST

1. Simplify the following expressions:

   a. $a + a + a + a$

   b. $5a + 2b + 3a - b$

   c. $6a - 3b + 2a - 4b$

   d. $12xy + 4xy - xy$

   e. $3a^2 - 6b^2 - 2b^2 + a^2$

   f. $5xy - 3yx + 2xy^2$

EXAM PRACTICE

1. Simplify the following expressions:

   a. $5ab - 2bc + 6bc - 7ab$

   b. $d \times d \times d \times d$

   c. $5m \times 3n$

2. Lauren buys x books costing £7 each and y magazines costing 98p each. Write down a formula for the total cost (T) of the books and magazines.

3. Adult cinema tickets cost £x and child cinema tickets cost £y. Mr Khan buys 2 adult tickets and 4 child tickets. Write down a formula in terms of x and y for the total cost (£C) of the tickets.

Formulae and Expressions 2

Substituting into Formulae

Replacing a letter with a number is called **substitution**.

Write out the expression first and then replace the letters with the values given.

Work out the value – but take care with the order of operations, i.e. BIDMAS.

Examples

1. $a = 3b - 4c$. Find a if $b = 4$ and $c = -2$

$a = (3 \times 4) - (4 \times -2)$
$ = 12 - (-8)$
$ = \mathbf{20}$

> Taking away a negative is the same as adding.

2. $E = \frac{1}{2}mv^2$. Find E if $m = 6$ and $v = 10$

$E = \frac{1}{2} \times 6 \times 10^2$
$ = \frac{1}{2} \times 6 \times 100$
$ = \mathbf{300}$

3. $V = u + at$. Find V if $u = 22$, $a = -2$ and $t = 6$

$V = 22 + (-2 \times 6)$
$ = 22 + (-12)$
$ = 22 - 12$
$ = \mathbf{10}$

4. $S = kt^2$. Find S when $k = 4$ and $t = -3$

$S = 4 \times (-3)^2$
$S = 4 \times 9$
$S = \mathbf{36}$

5. The cost of hiring a car can be worked out using this rule: Cost = £85 + 48p per mile.

a. Josh hires a car and drives 130 miles. Work out the cost.

$C = £85 + 48p \text{ per mile}$
$ = 85 + 0.48 \times 130$
$ = 85 + 62.40$
$ = \mathbf{£147.40}$

> Change 48p into £ so they are in the same unit: 48p = £0.48

b. On a different day Josh pays £131.56 for hiring the car. How many miles did he travel?

$131.56 = 85 + 0.48 \times \text{number of miles}$
$131.56 - 85 = 46.56$
$\dfrac{46.56}{0.48} = \mathbf{97 \text{ miles}}$

> Divide by the pence per mile to find the number of miles.

Rearranging Formulae

The **subject** of a formula is the letter that appears on its own on one side of the formula.

Any letter in a formula can become the subject by rearranging the formula.

Examples

1. Make c the subject of the formula:
   $$b = c - a$$

   $b = c - a$

   $b + a = c$ Add a to both sides.

   So $\boldsymbol{c = b + a}$

2. Make a the subject of the formula:
   $$b = (a - 3)^2$$

   $b = (a - 3)^2$ Deal with the power first. Square root both sides of the formula.

   $\pm\sqrt{b} = a - 3$ Remove any term added or subtracted. In this case add 3 to both sides of the formula.

   $\pm\sqrt{b} + 3 = a$ When square rooting we get a positive and negative solution, which is shown as $\pm$

   $\boldsymbol{a = \pm\sqrt{b} + 3}$

3. Make x the subject of the formula:
   $$5(y + x) = 8x + 3$$

   When the subject occurs on both sides of the equal sign, they need to be collected on one side.

   $5(y + x) = 8x + 3$

   $5y + 5x = 8x + 3$

   $5y - 3 = 8x - 5x$

   $5y - 3 = 3x$

   $$x = \frac{5y - 3}{3}$$

SUMMARY

- Substitution is replacing letters in a formula with numbers.

- Any letter can become the subject of a formula by rearranging the formula.

- When rearranging a formula, you must do the same thing to both sides of the formula.

QUESTIONS

QUICK TEST

1. 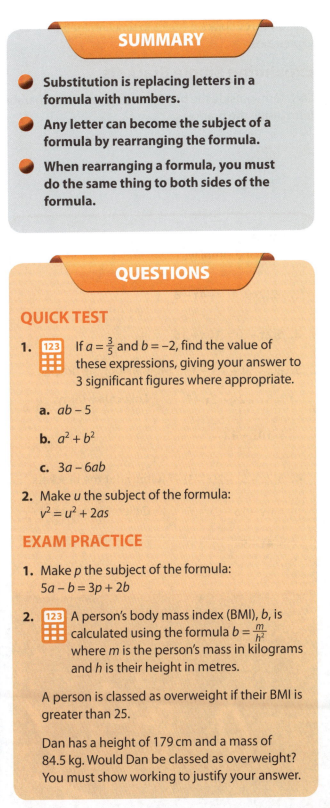 If $a = \frac{3}{5}$ and $b = -2$, find the value of these expressions, giving your answer to 3 significant figures where appropriate.

   a. $ab - 5$

   b. $a^2 + b^2$

   c. $3a - 6ab$

2. Make u the subject of the formula:
   $$v^2 = u^2 + 2as$$

EXAM PRACTICE

1. Make p the subject of the formula:
   $$5a - b = 3p + 2b$$

2. A person's body mass index (BMI), b, is calculated using the formula $b = \frac{m}{h^2}$ where m is the person's mass in kilograms and h is their height in metres.

   A person is classed as overweight if their BMI is greater than 25.

   Dan has a height of 179 cm and a mass of 84.5 kg. Would Dan be classed as overweight? You must show working to justify your answer.

Brackets and Factorisation

Multiplying out brackets helps to simplify algebraic expressions.

Expanding Single Brackets

Each term outside the bracket multiplies each separate term inside the bracket.

$$5(x + 6) = 5x + 30$$

Examples

Expand and simplify:

1. $-2(2x + 4) = \mathbf{-4x - 8}$

2. $5(2x - 3) = \mathbf{10x - 15}$

3. $8(x + 3) + 2(x - 1)$ Multiply out the brackets.

   $= 8x + 24 + 2x - 2$ Collect like terms.

   $= \mathbf{10x + 22}$

4. $3(2x - 5) - 2(x - 3)$ Multiply out the brackets.

   $= 6x - 15 - 2x + 6$ Collect like terms.

   $= \mathbf{4x - 9}$

Expanding Two Brackets

Every term in the second bracket must be multiplied by every term in the first bracket.

Often, but not always, the two middle terms are like terms and can be collected together.

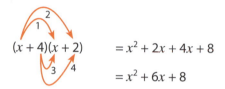

$$(x + 4)(x + 2) \quad = x^2 + 2x + 4x + 8$$
$$= x^2 + 6x + 8$$

Examples

Expand and simplify:

1. $(x + 4)(2x - 5) = 2x^2 - 5x + 8x - 20$

   $= \mathbf{2x^2 + 3x - 20}$

2. $(2x + 1)^2 = (2x + 1)(2x + 1)$

   $= 4x^2 + 2x + 2x + 1$

   $= \mathbf{4x^2 + 4x + 1}$

   Remember that x^2 means x multiplied by itself.

3. $(3x - 1)(x - 2) = 3x^2 - 6x - x + 2$

   $= \mathbf{3x^2 - 7x + 2}$

4. $(x - 4)(3x + 1) = 3x^2 + x - 12x - 4$

   $= \mathbf{3x^2 - 11x - 4}$

5. $(2x + 3y)(x - 2y) = 2x^2 - 4xy + 3xy - 6y^2$

   $= \mathbf{2x^2 - xy - 6y^2}$

Factorisation

Factorisation simply means putting an expression into brackets.

One Bracket

$4x + 6 = 2(2x + 3)$

To factorise $4x + 6$:

- Recognise that 2 is the Highest Common Factor of 4 and 6.

- Take out the common factor.

- The expression is completed inside the bracket so that when multiplied out it is equivalent to $4x + 6$.

Two Brackets

Two brackets are obtained when a quadratic expression of the type $ax^2 + bx + c$ is factorised.

> **Examples**
> 1. $x^2 + 4x + 3 = (x + 1)(x + 3)$
>
> 2. $x^2 - 7x + 12 = (x - 3)(x - 4)$
>
> 3. $x^2 + 3x - 10 = (x + 5)(x - 2)$
>
> 4. $x^2 - 64 = (x - 8)(x + 8)$
>
> This is known as the 'difference of two squares'. In general, $x^2 - a^2 = (x - a)(x + a)$.
>
> 5. $81x^2 - 25y^2 = (9x - 5y)(9x + 5y)$

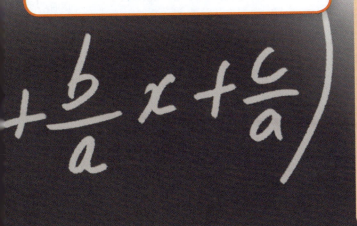

QUESTIONS

QUICK TEST

1. Expand and simplify:

   a. $(x + 3)(x - 2)$

   b. $4x(x - 3)$

   c. $(x - 3)^2$

2. Factorise:

   a. $12xy - 6x^2$

   b. $3a^2b + 6ab^2$

   c. $x^2 + 4x + 4$

   d. $x^2 - 4x - 5$

   e. $x^2 - 100$

EXAM PRACTICE

1. Expand and simplify:

   a. $t(3t - 4)$

   b. $4(2x - 1) - 2(x - 4)$

2. Factorise:

   a. $y^2 + y$

   b. $5p^2q - 10pq^2$

   c. $(a + b)^2 + 4(a + b)$

   d. $x^2 - 5x + 6$

Equations 1

Equations involve an unknown value that needs to be worked out.

Equations need to be kept balanced, so whatever is done to one side of the equation (for example, adding) also needs to be done to the other side.

Linear Equations of the Form $ax + b = c$

Examples

1. Solve: $3x = 15$

$x = \dfrac{15}{3}$ ← Divide both sides by 3.

$x = \mathbf{5}$

2. Solve: $\dfrac{x}{3} = 6$

$x = 6 \times 3$ ← Multiply both sides by 3.

$x = \mathbf{18}$

3. Solve: $5x - 2 = 13$

$5x = 13 + 2$ ← Add 2 to both sides.

$5x = 15$

$x = \dfrac{15}{5}$ ← Divide both sides by 5.

$x = \mathbf{3}$

4. Solve: $3x + 1 = 13$

$3x = 13 - 1$ ← Subtract 1 from both sides.

$3x = 12$

$x = \dfrac{12}{3}$ ← Divide both sides by 3.

$x = \mathbf{4}$

5. Solve: $\dfrac{x}{6} - 1 = 3$

$\dfrac{x}{6} = 3 + 1$ ← Add 1 to both sides.

$\dfrac{x}{6} = 4$

$x = 4 \times 6$ ← Multiply both sides by 6.

$x = \mathbf{24}$

Linear Equations of the Form $ax + b = cx + d$

Examples

1. Solve:

$7x - 4 = 3x + 8$

$7x = 3x + 12$ ← Add 4 to both sides.

$4x = 12$

$x = \dfrac{12}{4}$ ← Subtract $3x$ from both sides.

$x = \mathbf{3}$

Check by substituting 3 into both sides of the equation:

$7 \times 3 - 4 = 17$

$3 \times 3 + 8 = 17$

Since both the left-hand side of the equation and the right-hand side of the equation give the same answer $x = 3$ is correct ✔

2. Solve:

$5x + 3 = 2x - 5$

$5x = 2x - 5 - 3$ ← Subtract 3 from both sides.

$5x = 2x - 8$

$5x - 2x = -8$ ← Subtract $2x$ from both sides.

$3x = -8$

$x = -\dfrac{8}{3}$

$x = \mathbf{-2\dfrac{2}{3}}$

Linear Equations with Brackets

Examples

1. Solve:

$$5(x - 1) = 3(x + 2)$$

$$5x - 5 = 3x + 6$$

$$5x = 3x + 11$$

$$2x = 11$$

$$x = \frac{11}{2}$$

$$x = \mathbf{5.5}$$

> Just multiply out the brackets, then solve as normal.

2. Solve:

$$5(2x + 3) = 2(x - 6)$$

$$10x + 15 = 2x - 12$$

$$10x = 2x - 12 - 15$$

$$10x = 2x - 27$$

$$10x - 2x = -27$$

$$8x = -27$$

$$x = -\frac{27}{8}$$

$$x = \mathbf{-3\frac{3}{8}}$$

3. Solve:

$$\frac{3(2x - 1)}{5} = 6$$

$$3(2x - 1) = 6 \times 5$$

> Multiply both sides by 5.

$$6x - 3 = 30$$

$$6x = 33$$

$$x = \frac{33}{6}$$

$$x = \mathbf{5.5}$$

SUMMARY

- Whatever you do to one side of the equation needs to be done to the other side of the equation as well.
- Work through the solution step by step.
- Check your solution by substituting in your answer.

QUESTIONS

QUICK TEST

Solve the following equations:

1. $2x - 6 = 10$

2. $5 - 3x = 20$

3. $4(2 - 2x) = 12$

4. $6x + 3 = 2x - 10$

5. $7x - 4 = 3x - 6$

6. $5(x + 1) = 3(2x - 4)$

EXAM PRACTICE

1. Solve the equations:

  a. $5x - 3 = 9$

  b. $7x + 4 = 3x - 6$

  c. $3(4y - 1) = 21$

2. Solve:

  a. $5 - 2x = 3(x + 2)$

  b. $\frac{3x - 1}{3} = 4 + 2x$

Equations 2

Equation Problems

When solving equation problems, the first step is to write down the information that you know.

Example

The perimeter of this rectangle is 30 cm.

Work out the value of y and find the length of the rectangle.

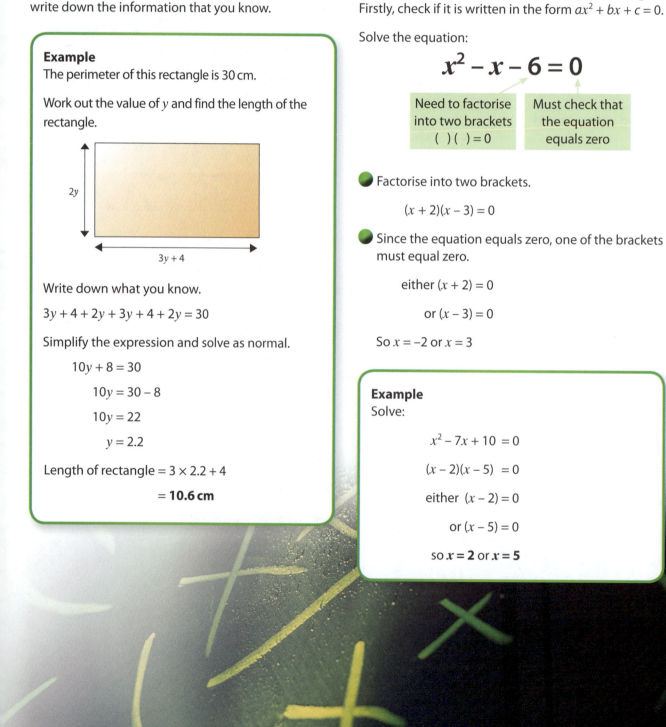

Write down what you know.

$3y + 4 + 2y + 3y + 4 + 2y = 30$

Simplify the expression and solve as normal.

$10y + 8 = 30$

$10y = 30 - 8$

$10y = 22$

$y = 2.2$

Length of rectangle $= 3 \times 2.2 + 4$

$= \mathbf{10.6\ cm}$

Solving Quadratic Equations

A **quadratic equation** can be solved by factorising. Firstly, check if it is written in the form $ax^2 + bx + c = 0$.

Solve the equation:

$$x^2 - x - 6 = 0$$

Need to factorise into two brackets ( )( ) = 0	Must check that the equation equals zero

● Factorise into two brackets.

$(x + 2)(x - 3) = 0$

● Since the equation equals zero, one of the brackets must equal zero.

either $(x + 2) = 0$

or $(x - 3) = 0$

So $x = -2$ or $x = 3$

Example

Solve:

$x^2 - 7x + 10 = 0$

$(x - 2)(x - 5) = 0$

either $(x - 2) = 0$

or $(x - 5) = 0$

so $x = \mathbf{2}$ or $x = \mathbf{5}$

Solving Cubic and Other Equations

Trial and improvement gives an approximate solution to cubic and other equations.

Example

The equation $x^3 + 2x = 58$ has a solution between 3 and 4. Find the solution to 1 decimal place.

Drawing a table can help you and the examiner, since it makes it easier to follow what you have done.

x	$x^3 + 2x$	Comment
3.5	$3.5^3 + 2 \times 3.5 = 49.875$	too small
3.8	$3.8^3 + 2 \times 3.8 = 62.472$	too big
3.7	$3.7^3 + 2 \times 3.7 = 58.053$	too big
3.6	$3.6^3 + 2 \times 3.6 = 53.856$	too small
3.65	$3.65^3 + 2 \times 3.65 = 55.927$	too small

3.5	3.6	3.65	3.7	3.8
too small	too small	too small	too big	too big

So $x = $ **3.7** (1dp) | Since the exact value of x is between 3.65 and 3.7

SUMMARY

- When solving equation problems, write down what you know, simplify and then solve as normal.

- Quadratic equations can be solved by factorisation.

- Cubic equations and other equations can be solved by trial and improvement.

QUESTIONS

QUICK TEST

1. The perimeter of this triangle is 60 cm. Work out the value of x and find the shortest length.

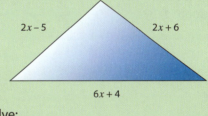

2. Solve:
   a. $x^2 - 7x = 0$ **b.** $x^2 + 8x + 15 = 0$
   c. $x^2 - 5x + 6 = 0$

EXAM PRACTICE

1. The sizes of the angles, in degrees, of the quadrilateral are:

   $x + 30°$
   $2x$
   $x + 50°$
   $x + 10°$

   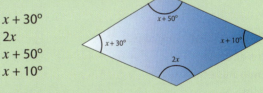

   Work out the smallest angle of the quadrilateral.

2. [123] The equation
   $x^3 + 4x^2 = 49$
   has a solution between $x = 2$ and $x = 3$.

   Use a trial and improvement method to find this solution. Give your answer correct to 1 decimal place. You must show all your working.

Simultaneous Linear Equations

Two equations with two unknowns are called **simultaneous equations**. They can be solved algebraically or graphically.

Solving Algebraically (Elimination Method)

Solve simultaneously:
$$3x + 2y = 8$$
$$2x - 3y = 14$$

Label the equations ① and ②.
$$3x + 2y = 8 \quad ①$$
$$2x - 3y = 14 \quad ②$$

Since no coefficients match, multiply equation ① by 2 and equation ② by 3.
$$6x + 4y = 16$$
$$6x - 9y = 42$$

Rename them equations ③ and ④.
$$6x + 4y = 16 \quad ③$$
$$6x - 9y = 42 \quad ④$$

The coefficient of x in equations ③ and ④ is the same. Subtract equation ④ from equation ③ and solve to find y.
$$0x + 13y = -26$$
$$y = -26 \div 13$$
$$y = -2$$

Note $4y - (-9y)$
$= 4y + 9y$
$= 13y$

Substitute the value of $y = -2$ into equation ①. Solve this equation to find x.
$$3x + 2 \times (-2) = 8$$
$$3x + (-4) = 8$$
$$3x = 8 + 4$$
$$3x = 12$$
$$x = 4$$

You could substitute into ②

Check in equation ②.
$$(2 \times 4) - (3 \times -2) = 14 \; ✔$$

Solution is: $x = 4$, $y = -2$

Solving Graphically

The point at which any two graphs intersect represents the simultaneous solutions of their equations.

Example

Solve the simultaneous equations:

$2x + 3y = 6$

$x + y = 1$

Draw the graph of:

$2x + 3y = 6$

When $x = 0$, $3y = 6$ $\therefore$ $y = 2$ (0, 2)

When $y = 0$, $2x = 6$ $\therefore$ $x = 3$ (3, 0)

Draw the graph of:

$x + y = 1$

When $x = 0$, $y = 1$ (0, 1)

When $y = 0$, $x = 1$ (1, 0)

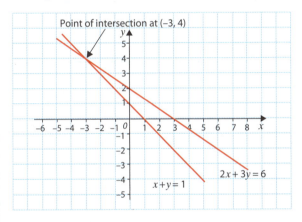

At the point of intersection: $x = -3$, $y = 4$

This is the solution of the simultaneous equations.

SUMMARY

- **When you are solving simultaneous equations, there are two equations with two unknowns.**

- **The elimination method involves eliminating one of the variables.**

- **The point where two graphs cross gives the solution to both equations.**

QUESTIONS

QUICK TEST

1. Solve the simultaneous equations:

   $4b + 7a = 10$

   $2b + 3a = 3$

2. The diagram shows the graphs of the lines:

   $x + y = 6$ and $y = x + 2$

   Use the diagram to solve the simultaneous equations $x + y = 6$ and $y = x + 2$

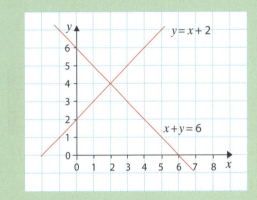

EXAM PRACTICE

1. Solve the simultaneous equations:

   $5a - 2b = 19$

   $3a + 4b = 1$

2. Frances and Patrick were organising a children's party. They went to the toy shop and bought some hats and balloons. Frances bought 5 hats and 4 balloons. She paid £22. Patrick bought 3 hats and 5 balloons. He paid £21.

   The cost of a hat was x pounds. The cost of a balloon was y pounds. Work out the cost of one hat and the cost of one balloon.

Sequences

A **sequence** is a set of numbers that follow a particular rule. The word '**term**' is often used to describe the numbers in the sequence.

Special Sequences

Odd numbers	$1, 3, 5, 7, 9 \dots$	nth term is $2n - 1$
Even numbers	$2, 4, 6, 8, 10 \dots$	nth term is $2n$

Square Numbers

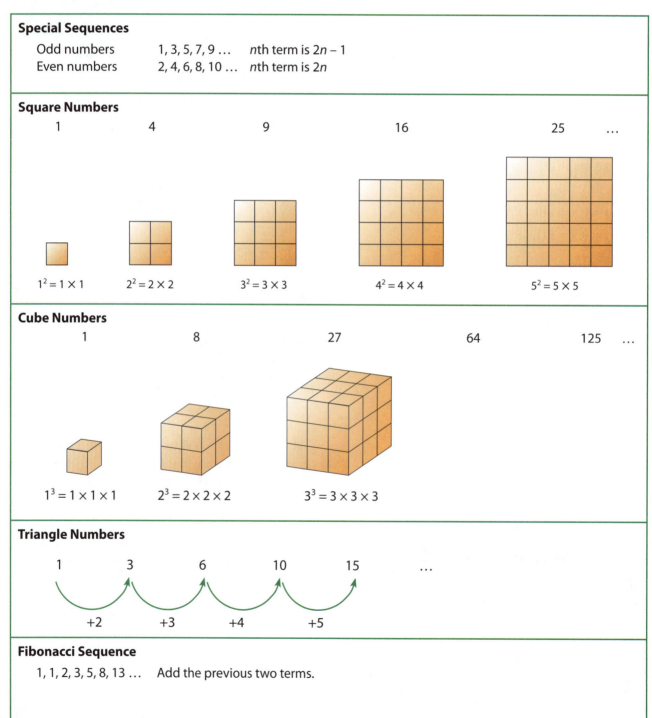

1	4	9	16	25	...

$1^2 = 1 \times 1$ $2^2 = 2 \times 2$ $3^2 = 3 \times 3$ $4^2 = 4 \times 4$ $5^2 = 5 \times 5$

Cube Numbers

1	8	27	64	125	...

$1^3 = 1 \times 1 \times 1$ $2^3 = 2 \times 2 \times 2$ $3^3 = 3 \times 3 \times 3$

Triangle Numbers

1   3   6   10   15   ...

+2   +3   +4   +5

Fibonacci Sequence

$1, 1, 2, 3, 5, 8, 13 \dots$ Add the previous two terms.

Finding the *n*th Term

The *n*th term is the rule for a sequence and is often denoted by U_n. For example, the 8th term is U_8.

For a linear sequence the *n*th term takes the form:

$$U_n = an + b$$

Example
Find the *n*th term of this sequence: 2, 6, 10, 14.

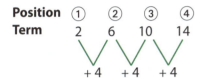

Position ① ② ③ ④
Term 2 6 10 14
+ 4 + 4 + 4

● See how the numbers are jumping (going up in 4s).

● The *n*th term is 4*n* + or – something.

● Try out 4*n* on the first term. This gives $4 \times 1 = 4$, but the first term is 2 … so subtract 2.

● The rule is 4*n* – 2

● Test this rule on the other terms:

  $1 \rightarrow (4 \times 1) - 2 = 2$

  $2 \rightarrow (4 \times 2) - 2 = 6$

  $3 \rightarrow (4 \times 3) - 2 = 10$

  It works on all of them.

● *n*th term is 4*n* – 2

  The 20th term in the sequence would be:

  $(4 \times 20) - 2 = 78$

SUMMARY

● A sequence is a set of numbers that follow a particular rule.

● The *n*th term is the rule for a sequence and is denoted by U_n.

● Given a sequence, you will need to be able to work out the *n*th term.

QUESTIONS

QUICK TEST

1. The cards show the *n*th term of some sequences:

   | 2*n* | 4*n* + 1 | 3*n* + 2 | 5*n* – 1 | 2 – *n* |

   Match the cards with the sequences below:

   a. 5, 9, 13, 17 …

   b. 1, 0, –1, –2 …

   c. 2, 4, 6, 8, 10 …

   d. 5, 8, 11, 14, 17 …

   e. 4, 9, 14, 19 …

EXAM PRACTICE

1. **a.** Here are the first four terms of an arithmetic sequence:

   5 7 9 11

   Find an expression in terms of *n* for the *n*th term of the sequence.

   b. The *n*th term of a different sequence is $2n^2 + 1$. Chloe says that 101 is a number in the sequence. Explain whether Chloe is correct.

Inequalities

In **inequalities** the left-hand side does not equal the right-hand side.

Inequalities can be solved in exactly the same way as equations except that when multiplying or dividing by a negative number, you must reverse the inequality sign.

The Inequality Symbols

$>$ means **greater than**

$<$ means **less than**

$\geqslant$ means **greater than or equal to**

$\leqslant$ means **less than or equal to**

Examples

1. Solve:

$$2x - 2 < 10$$
$$2x < 10 + 2$$
$$2x < 12$$
$$\boldsymbol{x < 6}$$

2. Solve:

$$3 - 2x \geqslant 9$$
$$-2x \geqslant 9 - 3$$
$$-2x \geqslant 6$$
$$x \leqslant \frac{6}{-2}$$
$$\boldsymbol{x \leqslant -3}$$

Divide by –2 and reverse the inequality.

3. Solve:

$$-7 < 3x - 1 \leqslant 11$$
$$-6 < 3x \leqslant 12$$
$$\boldsymbol{-2 < x \leqslant 4}$$

Add 1 to each part of the inequality.

Divide each part of the inequality by 3.

The integer values that satisfy this inequality are –1, 0, 1, 2, 3, 4.

Number Lines

Inequalities can be shown on a number line.

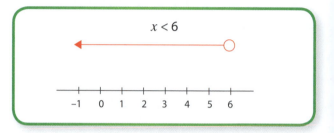

$$x < 6$$

The open circle means that 6 is not included.

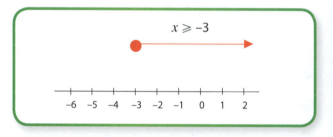

$$x \geqslant -3$$

The solid circle means that –3 is included.

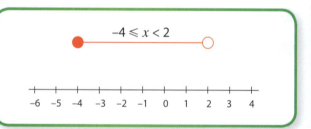

$$-4 \leqslant x < 2$$

The integer values that satisfy this inequality are –4, –3, –2, –1, 0, 1.

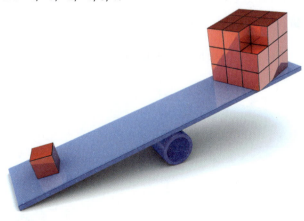

Graphs of Inequalities

The graph of an equation such as $x = 2$ is a line, whereas the graph of the inequality $x < 2$ is a region that has $x = 2$ as its boundary.

The diagram below shows unshaded the region (R):

$x + y \leqslant 7$

$x > 2$

$y \geqslant 2$

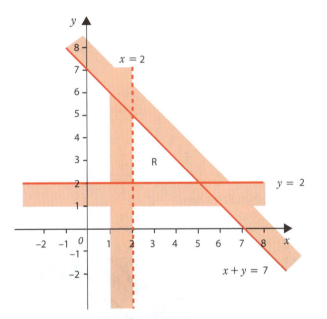

For strict inequalities $>$ and $<$ the boundary line is not included and is shown as a dashed line.

SUMMARY

- $>$ greater than
  $<$ less than
  $\geqslant$ greater than or equal to
  $\leqslant$ less than or equal to

- On number lines, an open circle means the value is not included in the inequality and a solid circle means the value is included in the inequality.

- On graphs of inequalities, use a dashed line when the boundary is not included.

QUESTIONS

QUICK TEST

1. Solve the following inequalities:

   a. $5x - 1 < 10$

   b. $6 \leqslant 3x + 2 < 11$

   c. $3 - 5x < 12$

EXAM PRACTICE

1. a. n is an integer such that $-6 < 2n \leqslant 8$.

   List all the possible values of n.

   b. Solve the inequality $4 + x > 7x - 8$.

2. On the diagram below, leave unshaded the region satisfied by these inequalities:

   $x + y \leqslant 5$
   $x \geqslant 1$
   $y > 1$

   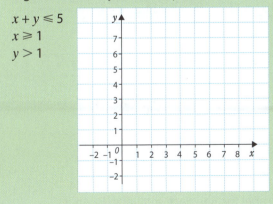

Straight-line Graphs

Drawing Straight-line Graphs

The general equation of a straight line is:

$$y = mx + c$$

m is the gradient. c is the intercept on the y-axis.

To draw the graph of $y = 3x - 4$:

🟢 Work out the coordinates of the points that lie on the line $y = 3x - 4$, by drawing a table of values of x.

Substitute the x values into the equation $y = 3x - 4$, to find the values of y
e.g. $x = 2$, $y = 3 \times 2 - 4 = 2$

x	−1	0	2	4
y	−7	−4	2	8

🟢 The coordinates of the points on the line are:

(−1, −7)   (0, −4)   (2, 2)   (4, 8)

Just read them from the table of values.

🟢 Plot the points (across the whole grid) and join with a straight line.

🟢 The line $y = 3x - 4$ is drawn.

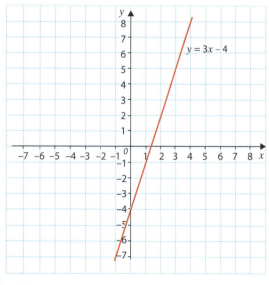

🟢 Label the line once you have drawn it.

Gradient of a Straight Line

Be careful when finding the gradient: double-check the scales.

$$\text{Gradient} = \frac{\text{change in } y}{\text{change in } x}$$

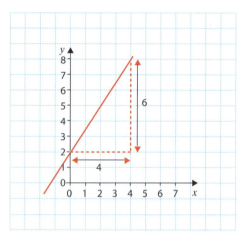

For the straight line above:

$$\text{Gradient} = \frac{6}{4} = \frac{3}{2} = 1.5$$

Positive and Negative Gradients

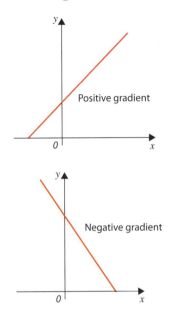

Positive gradient

Negative gradient

Finding the Midpoint of a Line Segment

The midpoint of a line segment between two points can be found by finding the mean of the x coordinates and the mean of the y coordinates of the points.

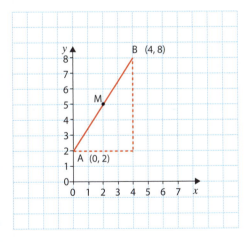

The midpoint of a line that joins the point $A(x_1, y_1)$ and $B(x_1, y_1)$ is:

$$\left(\frac{(x_1 + x_2)}{2}, \frac{(y_1 + y_2)}{2}\right)$$

The midpoint, M, of the line AB drawn here is:

$$\left(\frac{(0 + 4)}{2}, \frac{(2 + 8)}{2}\right) = (2, 5)$$

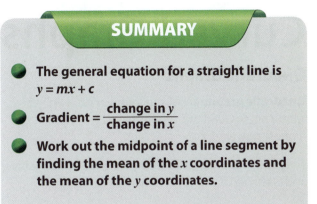

SUMMARY

- The general equation for a straight line is $y = mx + c$
- Gradient $= \dfrac{\text{change in } y}{\text{change in } x}$
- Work out the midpoint of a line segment by finding the mean of the x coordinates and the mean of the y coordinates.

QUESTIONS

QUICK TEST

1. a. Complete the table of values for $y = 2x + 3$.

x	−2	−1	0	1	2	3
y						

b. Draw the graph of $y = 2x + 3$.

EXAM PRACTICE

1.

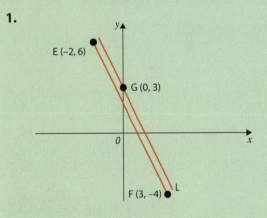

The diagram shows three points:

E (−2, 6)   F (3, −4)   G (0, 3)

A line L is parallel to EF and passes through G.

a. Find an equation for the line L.

b. Find the midpoint of the line EF.

Curved Graphs

Quadratic Graphs

Quadratic graphs are of the form $y = ax^2 + bx + c$ where $a \neq 0$. Quadratic graphs have an x^2 term as the highest power of x.

They will be ∪ shaped if the coefficient of x^2 is positive, and ∩ shaped if the coefficient of x^2 is negative.

To draw the graph of $y = x^2 - 2x - 6$, using values of x from −2 to 4:

🟢 Draw a table of values.

Fill in the table of values by substituting the values of x into the equation.

e.g. $x = 1$, $y = 1^2 - 2 \times 1 - 6 = -7$
Coordinates are $(1, -7)$

x	−2	−1	0	1	2	3	4
y	2	−3	−6	−7	−6	−3	2

🟢 Draw the axes on graph paper and plot the points.

🟢 Join the points with a smooth curve.

🟢 Label the curve.

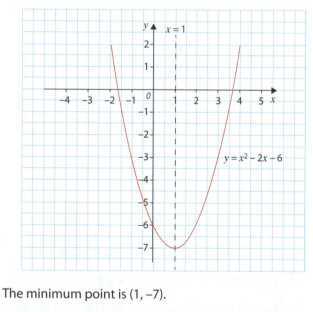

The minimum point is $(1, -7)$.

The line of symmetry is $x = 1$.

Graph Shapes

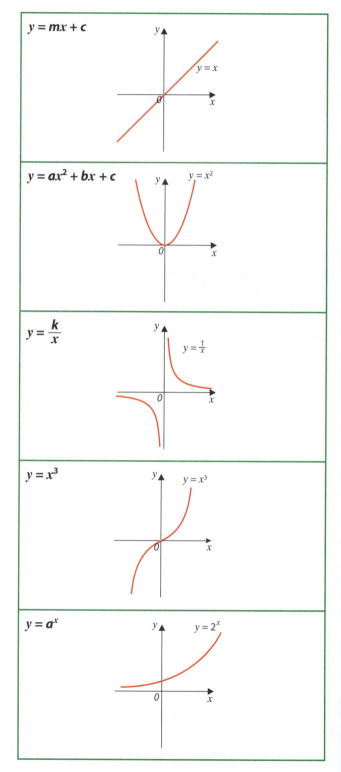

Example
Match each graph to one of the following equations.

$y = x^2 - 4$ $y = 5 - 2x$ $y = x^3$ $y = 3x - 1$

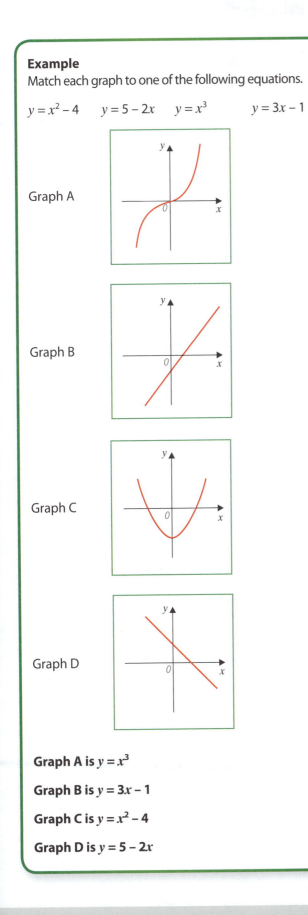

Graph A

Graph B

Graph C

Graph D

Graph A is $y = x^3$

Graph B is $y = 3x - 1$

Graph C is $y = x^2 - 4$

Graph D is $y = 5 - 2x$

SUMMARY

● **Make sure you know the different graph shapes.**

● **Quadratic graphs have an x^2 term as the highest power of x. They are ∪ shaped or ∩ shaped.**

● **Cubic graphs have an x^3 term as the highest power of x.**

● **The reciprocal function is $y = \dfrac{k}{x}$**

QUESTIONS

QUICK TEST

1. Match each graph below to one of the equations.

$y = x^3 - 5$ $y = 2 - x^2$

$y = 4x + 2$ $y = \dfrac{3}{x}$

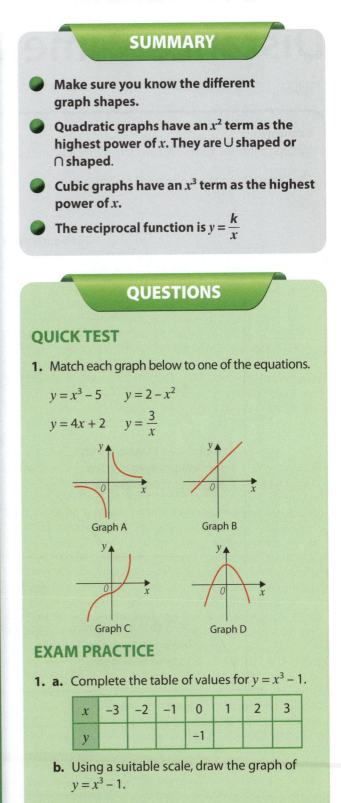

Graph A   Graph B

Graph C   Graph D

EXAM PRACTICE

1. **a.** Complete the table of values for $y = x^3 - 1$.

x	−3	−2	−1	0	1	2	3
y			−1				

b. Using a suitable scale, draw the graph of $y = x^3 - 1$.

c. From the graph find the approximate value of x when $y = 15$.

Distance–Time Graphs

Distance–time graphs are often known as travel graphs.

Example

Mr Smith travels from St Albans to his office 80 miles away.

The distance–time graph shows his journey.

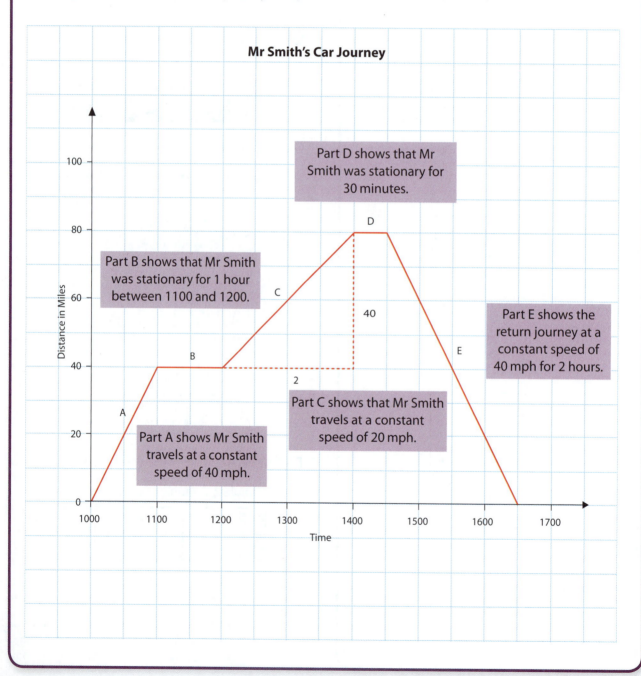

Mr Smith's Car Journey

Part D shows that Mr Smith was stationary for 30 minutes.

Part B shows that Mr Smith was stationary for 1 hour between 1100 and 1200.

Part E shows the return journey at a constant speed of 40 mph for 2 hours.

Part C shows that Mr Smith travels at a constant speed of 20 mph.

Part A shows Mr Smith travels at a constant speed of 40 mph.

Distance in Miles

Time

Always check that you understand the scales on a distance–time graph. For this graph:

● vertically 1 square represents 10 miles

● horizontally 2 squares represents 1 hour

The gradient of the distance–time graph represents the speed over that time interval.

$$\text{Speed} = \frac{\text{distance travelled}}{\text{time taken}}$$

In part C of the graph:

$$\text{speed} = \frac{40}{2}$$

$$\text{speed} = 20 \text{ mph}$$

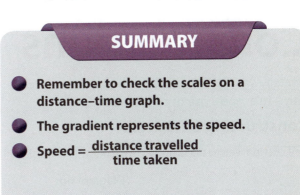

SUMMARY

● **Remember to check the scales on a distance–time graph.**

● **The gradient represents the speed.**

● **Speed = $\dfrac{\text{distance travelled}}{\text{time taken}}$**

QUESTIONS

QUICK TEST

1. 🔢 A car travels 120 kilometres in 2.5 hours. What is the speed of the car?

2. 🔢 John walks 4.5 miles in 1.5 hours. Work out his speed.

EXAM PRACTICE

1. The distance–time graph shows the car journeys of two people. Decide whether these statements are true or false.

   a. Miss Roberts travelled the first 100 miles at a speed of 50 mph.

   b. Mr Cohen had a rest for 1 hour between 0900 and 1000.

   c. Mr Cohen travelled at a speed of 37.5 mph between 1000 and 1200.

   d. Miss Roberts and Mr Cohen pass each other at approximately 0820.

   e. Miss Roberts travelled at a speed of 50 mph between 0930 and 1030.

Constructions

The following constructions can be completed using only a ruler and a pair of compasses.

Constructing a Triangle

Use compasses to accurately construct this triangle.

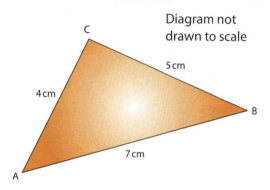

Diagram not drawn to scale

- Draw the longest side AB.

- With the compass point at A, draw an arc of radius 4 cm.

- With the compass point at B, draw an arc of radius 5 cm.

- Join A and B to the point where the two arcs meet at C.

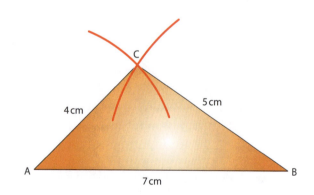

Remember to leave in your arcs. No arcs, no marks!

The Perpendicular Bisector of a Line

- Draw a line XY.

- Draw two arcs with the compasses, using X as the centre. The compasses must be set at a radius greater than half the distance of XY.

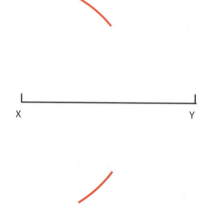

- Draw two more arcs with Y as the centre. (Keep the compasses the same distance apart as before.)

- Join the two points where the arcs cross.

- AB is the **perpendicular bisector** of XY.

- N is the **midpoint** of XY.

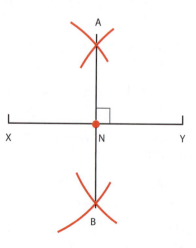

The Perpendicular from a Point to a Line

⬤ From P draw arcs to cut the line at A and B.

⬤ From A and B draw arcs with the same radius to intersect at C.

⬤ Join P to C; this line is perpendicular to AB.

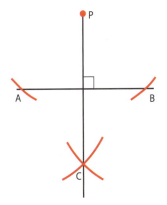

Bisecting an Angle

⬤ Draw two lines XY and YZ to meet at an angle.

⬤ Using compasses, place the point at Y and draw arcs on XY and YZ.

⬤ Place the compass point at the two arcs on XY and YZ and draw arcs to cross at N. Join Y and N. YN is the bisector of angle XYZ.

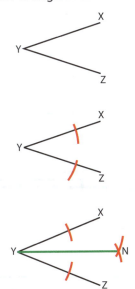

SUMMARY

⬤ **Always leave in your construction arcs.**

⬤ **Perpendicular means at right angles to.**

⬤ **Bisect means to cut in half.**

QUESTIONS

QUICK TEST

1. Construct an angle of 30°, using a ruler and pair of compasses only.

2. Draw the perpendicular bisector of an 8 cm line.

EXAM PRACTICE

1. **a.** Using compasses only, construct the perpendicular from the point P.
   You must show all construction lines.

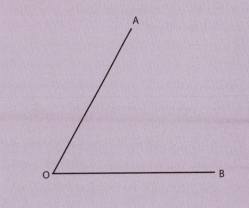

   b. Using compasses only, bisect this angle.

   You must show all construction lines.

Loci

The **locus** of a point is the set of all the possible positions that the point can occupy subject to some given condition or rule.

Types of Loci

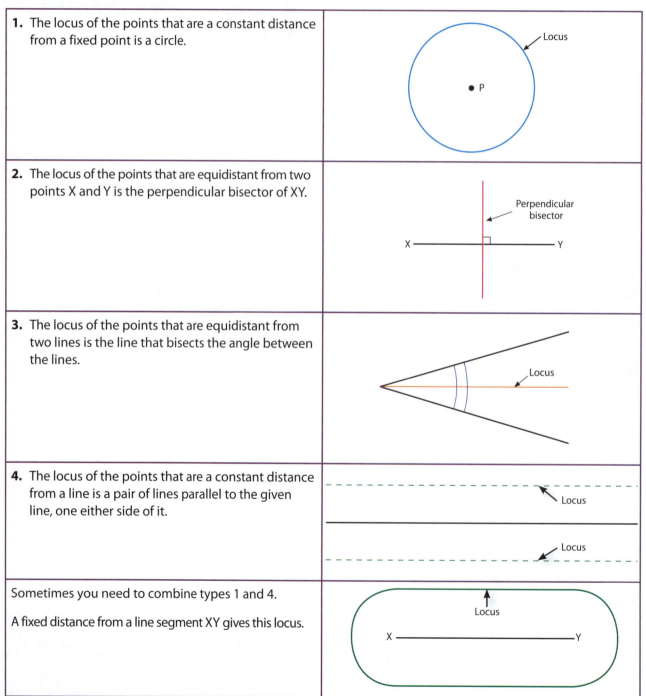

1. The locus of the points that are a constant distance from a fixed point is a circle.	
2. The locus of the points that are equidistant from two points X and Y is the perpendicular bisector of XY.	
3. The locus of the points that are equidistant from two lines is the line that bisects the angle between the lines.	
4. The locus of the points that are a constant distance from a line is a pair of lines parallel to the given line, one either side of it.	
Sometimes you need to combine types 1 and 4. A fixed distance from a line segment XY gives this locus.	

Example

Three radio transmitters form an equilateral triangle ABC with sides of 50 km. The range of the transmitter at A is 37.5 km, at B 30 km and at C 28 km. Using a scale of 1 cm to 10 km, construct a scale diagram to show where signals from all three transmitters can be received.

Below is a sketch not drawn to scale.

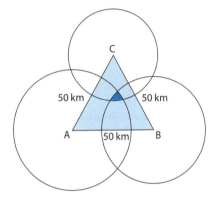

Please note that on your scale drawing the circle at A would have a radius of 3.75 cm. The circle at B would have a radius of 3 cm and the circle at C a radius of 2.8 cm.

The area where signals from all three transmitters can be received is shaded dark blue.

QUESTIONS

QUICK TEST

1. ABCD is a rectangle. The rectangle is accurately drawn.

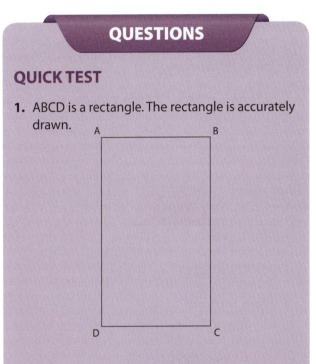

Shade the set of points inside the rectangle which are more than 2 cm from point B and more than 1.5 cm from the line AD.

EXAM PRACTICE

1. The plan shows a garden drawn to a scale of 1 cm : 2 m. A and B are bushes and C is a pond.

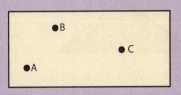

A landscape gardener has decided:

a. to lay a path right across the garden at an equal distance from each of the bushes, A and B.

b. to lay a flower border 2 m wide around pond C.

Construct these features on the plan above.

Angles

Angle Facts

Whenever lines meet or intersect, the angles they make follow certain rules:

1. Angles on a straight line add up to 180°.

$a + b + c = 180°$

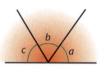

2. Angles at a point add up to 360°.

$a + b + c + d = 360°$

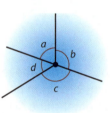

3. Angles in a triangle add up to 180°.

$a + b + c = 180°$

4. Vertically opposite angles are equal.

$a = b, c = d$

$a + d = b + c = 180°$

5. Angles in a quadrilateral add up to 360°.

$a + b + c + d = 360°$

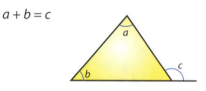

6. An exterior angle of a triangle equals the sum of the two opposite interior angles.

$a + b = c$

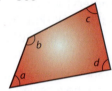

Examples
Find the missing angles in the diagrams below:

1.

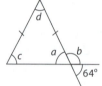

$a = \mathbf{64°}$ (opposite angles are equal)

$b = 180° - 64°$ (angles on a straight line add up to 180°)

$b = \mathbf{116°}$

$c = \mathbf{64°}$ (isosceles triangle base angles equal)

$d = \mathbf{52°}$ (angles in a triangle add up to 180°)

2.

$e = \mathbf{72°}$ (opposite angles are equal)

$f = \mathbf{108°}$ (angles on a straight line add up to 180°)

$g = \mathbf{108°}$

Parallel Lines

Three types of relationship are produced when a line called a transversal crosses a pair of parallel lines.

Alternate angles are equal.

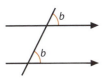

Corresponding angles are equal.

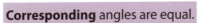

Supplementary angles add up to 180°. $c + d = 180°$

Angles in Polygons

There are two types of angle in a polygon – **interior** and **exterior**. A regular polygon has all sides and angles equal.

For a polygon with *n* sides:

- Sum of exterior angles = 360°

- Interior angle + exterior angle = 180°

- Sum of interior angles = $(n - 2) \times 180°$ or $(2n - 4) \times 90°$

For a regular polygon with *n* sides:

- Exterior angle = $\dfrac{360°}{n}$

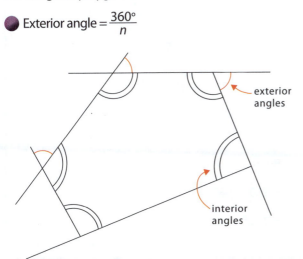

exterior angles

interior angles

SUMMARY

- **Angles add up to: 180° on a straight line, 360° at a point, 180° in a triangle, 360° in a quadrilateral.**

- **Alternate and corresponding angles are equal; supplementary angles add up to 180°.**

- **Polygons have interior and exterior angles.**

QUESTIONS

QUICK TEST

1. Work out the size of the angles in the diagrams below.

   a. **b.** **c.**

   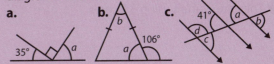

2. Work out the size of the exterior angle of a 12-sided regular polygon.

EXAM PRACTICE

1. BCD is a straight line. Explain why BE and CF must be parallel.

   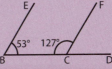

2. Work out the size of angle *y* in this polygon.

   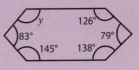

Bearings

Compass Directions

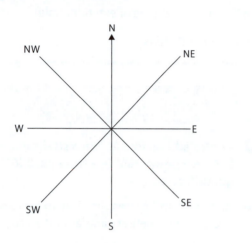

Bearings

Bearings are often used in scale drawing questions.

⬤ They are always measured from the North (N).

⬤ They are measured in a clockwise direction.

⬤ They are written using 3 figures.

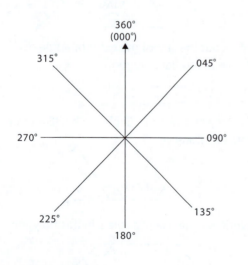

Examples

Find the bearings of P **from** Q in each diagram below:

a.

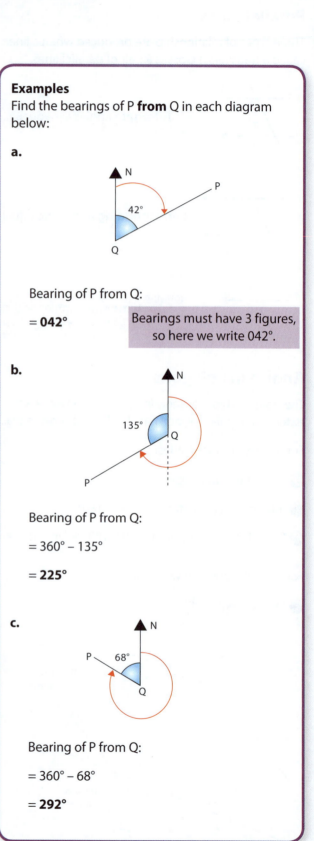

Bearing of P from Q:

= **042°**

> Bearings must have 3 figures, so here we write 042°.

b.

Bearing of P from Q:

= 360° – 135°

= **225°**

c.

Bearing of P from Q:

= 360° – 68°

= **292°**

Back Bearings

Back bearings are more difficult – you need to draw in a second North line. The angle properties of parallel lines can then be used.

Examples

Find the bearings of Q from P in each diagram below:

a.

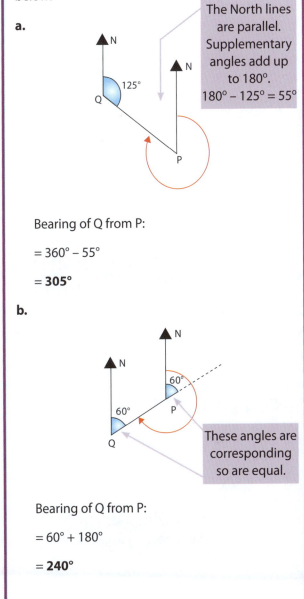

The North lines are parallel. Supplementary angles add up to 180°.
$180° - 125° = 55°$

Bearing of Q from P:

$= 360° - 55°$

$= 305°$

b.

Bearing of Q from P:

$= 60° + 180°$

$= 240°$

These angles are corresponding so are equal.

SUMMARY

- **Remember the key facts about bearings:**
    - **They are always measured from the North.**
    - **They are measured in a clockwise direction.**
    - **They are written using 3 figures, for example 060°.**

QUESTIONS

QUICK TEST

1. For the following diagrams find the bearing of B from A.

a.

$72°$

b.

$152°$

c.

$70°$

d.

$41°$

EXAM PRACTICE

1. The diagram shows the position of three towns A, B and C. Find the bearing of:

   a. A from B

   b. C from B

   $173°$ $98°$

2. A ship sails on a bearing of 143° to a buoy, A. Work out the bearing the ship needs to sail to return to its starting point from the buoy, A.

Translations and Reflections

Translations

Translations move figures from one position to another position. **Vectors** are used to describe the distance and direction of the translations.

A vector is written as $\binom{a}{b}$

a represents the horizontal distance and b represents the vertical distance.

The object and the image are **congruent** when the shape is translated, i.e. they are identical.

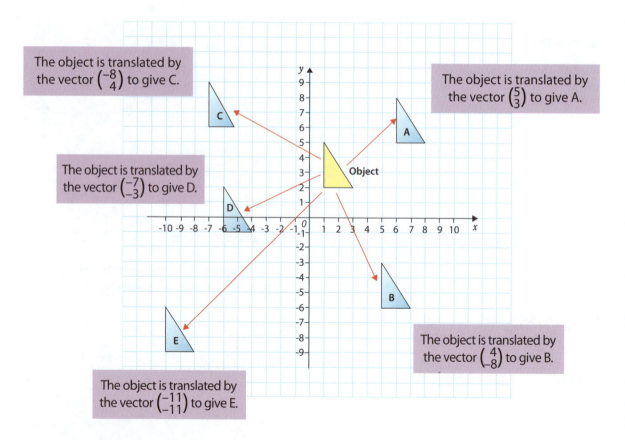

The object is translated by the vector $\binom{-8}{4}$ to give C.

The object is translated by the vector $\binom{5}{3}$ to give A.

The object is translated by the vector $\binom{-7}{-3}$ to give D.

The object is translated by the vector $\binom{4}{-8}$ to give B.

The object is translated by the vector $\binom{-11}{-11}$ to give E.

Reflections

Reflections create an image of an object on the other side of the mirror line.

The mirror line is known as an **axis of reflection**.

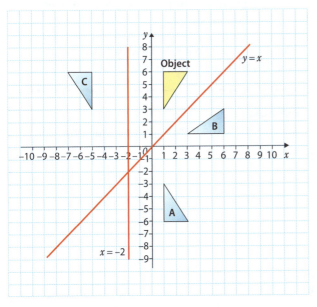

On the example above, the object is reflected in the x-axis (or $y = 0$) to give image A.

The object is reflected in the line $y = x$ to give image B.

The object is reflected in the line $x = -2$ to give image C.

The images and object are congruent.

QUESTIONS

QUICK TEST

1. For the diagram below, describe fully the transformation that maps:

   a. A onto B   **b.** B onto C

   c. A onto D   **d.** A onto E

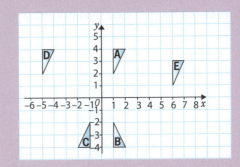

EXAM PRACTICE

1. Triangle A and triangle B have been drawn on the grid.

   a. Reflect triangle A in the line $x = 5$
      Label this image C.

   b. Translate triangle B by the vector $\begin{pmatrix} 4 \\ 2 \end{pmatrix}$
      Label this image D.

   c. Describe fully the single transformation which will map triangle A onto triangle B.

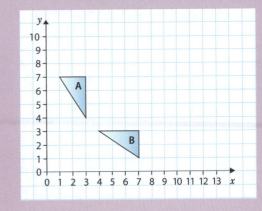

Rotation and Enlargement

Rotations

In a **rotation** the object is turned by a given angle about a fixed point called the **centre of rotation**. The size and shape of the figure are not changed, i.e. the image is **congruent** to the object.

On the example below, object A is rotated by 90° clockwise about (0, 0) to give image B.

Object A is rotated by 180° about (0, 0) to give image C.

Object A is rotated 90° anticlockwise about (−2, 2) to give image D.

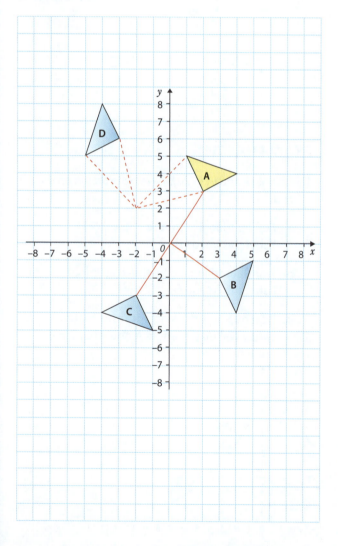

Enlargements

Enlargements change the size but not the shape of the object, i.e. the enlarged shape is **similar** to the object.

The **centre of enlargement** is the point from which the enlargement takes place.

The **scale factor** tells you what all lengths of the original figure have been multiplied by.

Example
Describe fully the transformation that maps ABC onto A′B′C′.

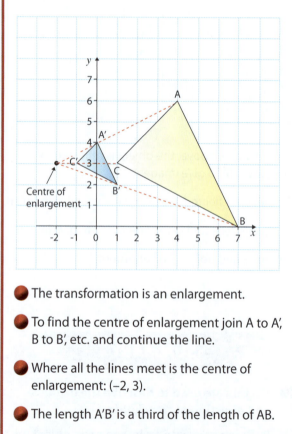

● The transformation is an enlargement.

● To find the centre of enlargement join A to A′, B to B′, etc. and continue the line.

● Where all the lines meet is the centre of enlargement: (−2, 3).

● The length A′B′ is a third of the length of AB.

The transformation is an enlargement by scale factor $\frac{1}{3}$. Centre of enlargement is (−2, 3).

An enlargement with a scale factor between 0 and 1 makes the shape smaller.

Enlargements with a Negative Scale Factor

For an enlargement with a **negative scale factor**, the image is situated on the opposite side of the centre of enlargement.

On the example below, the triangle A (3, 4), B (9, 4) and C (3, 10) is enlarged with a scale factor of $-\frac{1}{3}$, about the centre of enlargement (0, 1). The enlargement is labelled A'B'C'.

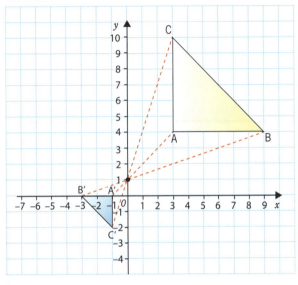

- A'B'C' is on the opposite side of the centre of enlargement from ABC.

- Notice that the length of each side of the triangle A'B'C' is one third the size of the corresponding lengths in the triangle ABC.

SUMMARY

- **After a rotation, the image is congruent to the object.**

- **After an enlargement, the enlarged shape is similar to the object.**

- **An enlargement with a scale factor between 0 and 1 makes the shape smaller.**

- **An enlargement with a negative scale factor means that the image is on the opposite side of the centre of enlargement.**

QUESTIONS

QUICK TEST

1. Complete the diagram below, to show the enlargement of the shape by a scale factor of $\frac{1}{2}$. Centre of enlargement at (0, 0). Call the shape T.

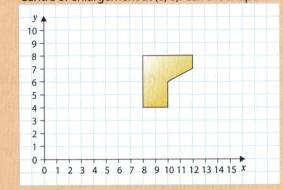

2. Rotate triangle ABC 90° anticlockwise about the point (0, 1). Call the triangle A'B'C'.

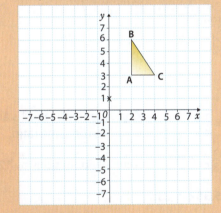

EXAM PRACTICE

1. The quadrilateral is enlarged with a scale factor of $-\frac{1}{2}$ about the origin (0, 0). Draw the enlargement on the diagram. Call the shape A'B'C'D'.

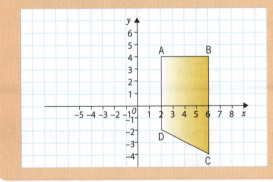

Similarity

Similar Shapes

Objects that are exactly the same shape but different sizes are called **similar** shapes. One is an enlargement of the other.

Corresponding angles are equal.

Corresponding lengths are in the same ratio.

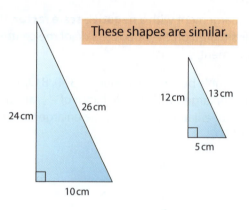

These shapes are similar.

Finding Missing Lengths of Similar Figures

Questions about finding missing lengths are very common at GCSE.

Examples

1. Find the missing lengths labelled a in the diagrams below:

a.

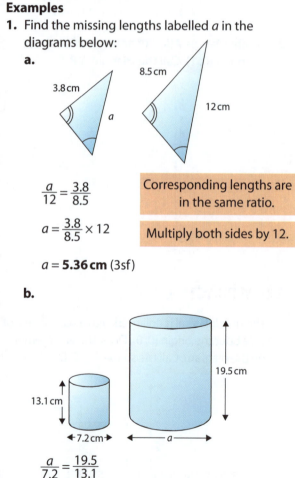

$$\frac{a}{12} = \frac{3.8}{8.5}$$

Corresponding lengths are in the same ratio.

$$a = \frac{3.8}{8.5} \times 12$$

Multiply both sides by 12.

$$a = \mathbf{5.36\,cm}\ (3sf)$$

b.

$$\frac{a}{7.2} = \frac{19.5}{13.1}$$

$$a = \frac{19.5}{13.1} \times 7.2$$

Multiply both sides by 7.2

$$a = \mathbf{10.7\,cm}\ (3sf)$$

2. Calculate the missing length y.

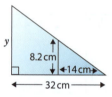

First draw out the individual triangles.

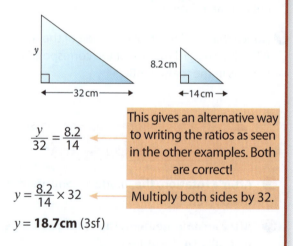

$$\frac{y}{32} = \frac{8.2}{14}$$

This gives an alternative way to writing the ratios as seen in the other examples. Both are correct!

$$y = \frac{8.2}{14} \times 32$$

Multiply both sides by 32.

$$y = \mathbf{18.7\,cm}\ (3sf)$$

More Difficult Problems

Sometimes you will be given a more difficult problem to solve.

Example

In the diagram CD is parallel to EF.

EF = 4.1 cm   FG = 5 cm   DG = 7.2 cm   CG = 12 cm

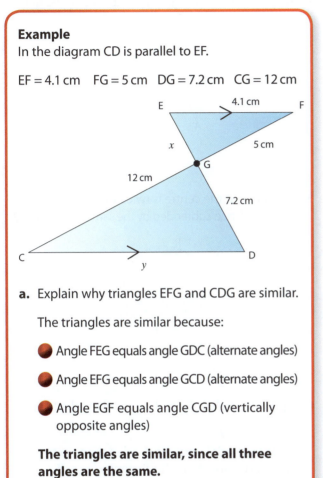

a. Explain why triangles EFG and CDG are similar.

The triangles are similar because:

- Angle FEG equals angle GDC (alternate angles)
- Angle EFG equals angle GCD (alternate angles)
- Angle EGF equals angle CGD (vertically opposite angles)

The triangles are similar, since all three angles are the same.

b. Calculate the lengths marked x and y.

To find x:

$$\frac{x}{5} = \frac{7.2}{12}$$

$$x = \frac{7.2}{12} \times 5$$

$$x = \textbf{3 cm}$$

To find y:

$$\frac{y}{12} = \frac{4.1}{5}$$

$$y = \frac{4.1}{5} \times 12$$

$$y = \textbf{9.84 cm}$$

SUMMARY

- Objects that are exactly the same shape but different sizes are called **similar shapes.**
- In similar shapes, corresponding angles are equal.
- In similar shapes, corresponding lengths are in the same ratio.

QUESTIONS

QUICK TEST

1. **123** Calculate the lengths marked n in these similar shapes. Give your answers correct to 1dp.

   a.

   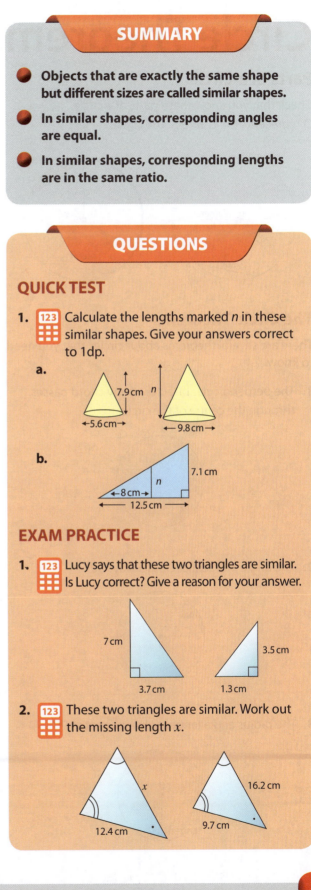

   7.9 cm n
   ←5.6 cm→   ←9.8 cm→

   b.

   7.1 cm
   n
   ←8 cm→
   12.5 cm

EXAM PRACTICE

1. **123** Lucy says that these two triangles are similar. Is Lucy correct? Give a reason for your answer.

   7 cm   3.5 cm
   3.7 cm   1.3 cm

2. **123** These two triangles are similar. Work out the missing length x.

   x 16.2 cm
   12.4 cm   9.7 cm

Circle Theorems

Parts of a Circle

Check that you know these parts of a circle.
O represents the centre of the circle.

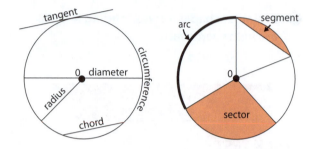

The Circle Theorems

There are several theorems about circles that you need to know.

1. The perpendicular bisector of any chord passes through the centre of the circle.

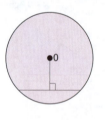

2. The angle in a semicircle is always 90°.

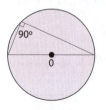

3. The radius and a tangent always meet at 90°.

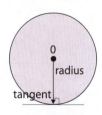

4. Angles in the same segment subtended by the same arc are equal, e.g. $A\hat{B}C = A\hat{D}C$.

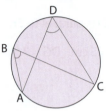

5. The angle at the centre is twice the angle at the circumference subtended by the same arc, e.g. $P\hat{O}Q = 2 \times P\hat{R}Q$.

6. Opposite angles of a cyclic quadrilateral add up to 180°.

(A cyclic quadrilateral is a 4-sided shape with each vertex touching the circumference of the circle.)

i.e. $x + y = 180°$

$a + b = 180°$

7. The lengths of two tangents from a point are equal, i.e. RS = RT.

Example

Calculate the angles marked *a–d* in the diagram below. Give reasons for your answers.

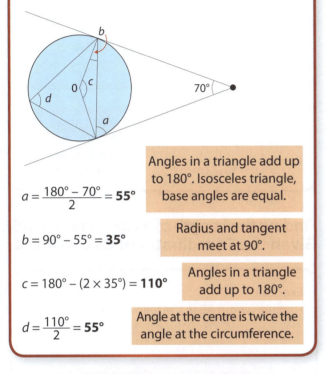

$a = \dfrac{180° - 70°}{2} = \textbf{55°}$

> Angles in a triangle add up to 180°. Isosceles triangle, base angles are equal.

$b = 90° - 55° = \textbf{35°}$

> Radius and tangent meet at 90°.

$c = 180° - (2 \times 35°) = \textbf{110°}$

> Angles in a triangle add up to 180°.

$d = \dfrac{110°}{2} = \textbf{55°}$

> Angle at the centre is twice the angle at the circumference.

SUMMARY

- **The angle between a tangent and radius of a circle is 90°.**

- **Tangents to a circle from a point outside the circle are equal in length.**

- **The angles in the same segment subtended by the same arc are equal.**

- **Opposite angles of a cyclic quadrilateral add up to 180°.**

- **The angle in a semicircle is a right angle.**

- **The angle at the centre of a circle is twice the angle at the circumference, both subtended by the same arc.**

QUESTIONS

QUICK TEST

1. Some angles are written on cards. Match the missing angles in the diagrams below with the correct card.

   0 represents the centre of the circle.

   | 53° | 50° | 62° | 109° | 126° |

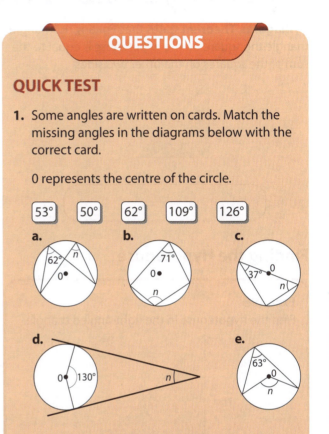

EXAM PRACTICE

1. In the diagram, E, F and G are points on the circle, centre 0.

   Angle FGB = 71°

   AB is a tangent to the circle at point G.

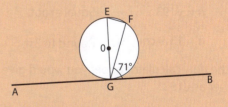

   a. Calculate the size of angle EGF. Give reasons for your answer.

   b. Calculate the size of angle GEF. Give reasons for your answer.

Pythagoras' Theorem

Pythagoras' Theorem states that 'For any right-angled triangle the square on the hypotenuse is equal to the sum of the squares on the other two sides.'

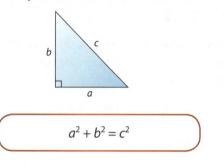

$$a^2 + b^2 = c^2$$

Finding the Hypotenuse

Example
Find the hypotenuse in the right-angled triangle.

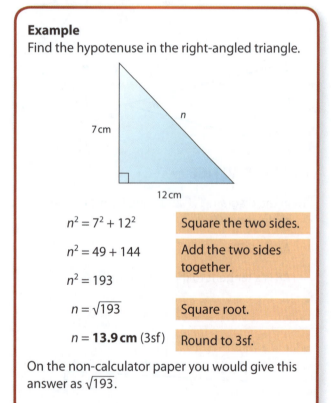

7 cm

12 cm

n

$n^2 = 7^2 + 12^2$	Square the two sides.
$n^2 = 49 + 144$	Add the two sides together.
$n^2 = 193$	
$n = \sqrt{193}$	Square root.
$n = \textbf{13.9 cm}$ (3sf)	Round to 3sf.

On the non-calculator paper you would give this answer as $\sqrt{193}$.

Finding a Short Side

Example
Find the length of p.

$$15^2 = p^2 + 8^2$$

$$15^2 - 8^2 = p^2$$

$$225 - 64 = p^2$$

$$161 = p^2$$

$$\sqrt{161} = p$$

$$p = \textbf{12.7 cm} \text{ (3sf)}$$

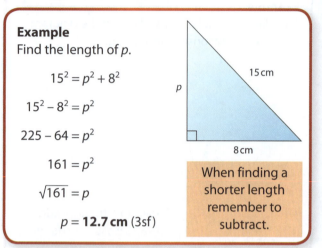

15 cm

p

8 cm

> When finding a shorter length remember to subtract.

Finding the Length of a Line Segment AB, Given the Coordinates of its End Points

Example
Find the length AB in the following diagram.

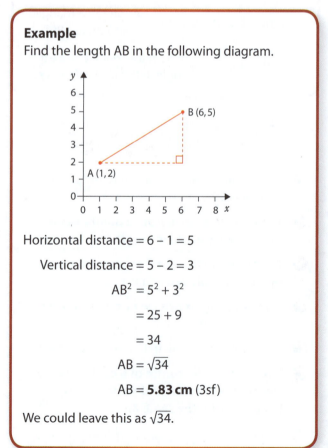

Horizontal distance = 6 – 1 = 5

Vertical distance = 5 – 2 = 3

$$AB^2 = 5^2 + 3^2$$

$$= 25 + 9$$

$$= 34$$

$$AB = \sqrt{34}$$

$$AB = \textbf{5.83 cm} \text{ (3sf)}$$

We could leave this as $\sqrt{34}$.

Solving a More Difficult Problem

Calculate the vertical height of this isosceles triangle.

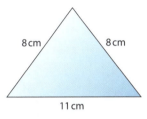

8 cm	8 cm
	11 cm

> Remember to split the triangle down the middle to make it right-angled.

Using Pythagoras' Theorem gives:

$$8^2 = h^2 + 5.5^2$$

$$64 = h^2 + 30.25$$

$$64 - 30.25 = h^2$$

$$33.75 = h^2$$

$$\sqrt{33.75} = h$$

$$h = 5.81 \text{ cm (3sf)}$$

8 cm h

5.5 cm

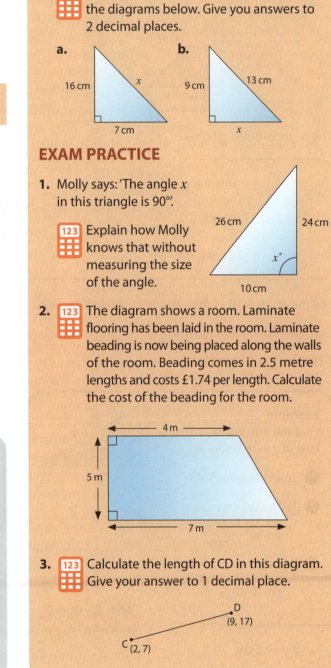

QUESTIONS

QUICK TEST

1. **123** Work out the missing lengths labelled x in the diagrams below. Give you answers to 2 decimal places.

   a.

   16 cm x

   7 cm

   b.

   9 cm 13 cm

   x

EXAM PRACTICE

1. Molly says: 'The angle x in this triangle is 90°'.

   123 Explain how Molly knows that without measuring the size of the angle.

   26 cm 24 cm

   $x°$

   10 cm

2. **123** The diagram shows a room. Laminate flooring has been laid in the room. Laminate beading is now being placed along the walls of the room. Beading comes in 2.5 metre lengths and costs £1.74 per length. Calculate the cost of the beading for the room.

   4 m

   5 m

   7 m

3. **123** Calculate the length of CD in this diagram. Give your answer to 1 decimal place.

   D (9, 17)

   C (2, 7)

Trigonometry

Trigonometry in right-angled triangles can be used to find an unknown angle or length.

The sides of a right-angled triangle are given temporary names according to where they are in relation to a chosen angle θ.

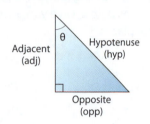

The trigonometric ratios are:

$$\text{Sine } \theta = \frac{\text{Opposite}}{\text{Hypotenuse}}$$

$$\text{Cosine } \theta = \frac{\text{Adjacent}}{\text{Hypotenuse}}$$

$$\text{Tangent } \theta = \frac{\text{Opposite}}{\text{Adjacent}}$$

Use the words **SOH – CAH – TOA** to remember the ratios.

Example: TOA means $\tan \theta = \dfrac{\text{opp}}{\text{adj}}$

Finding a Length

Example
Find the missing length y in the diagram.

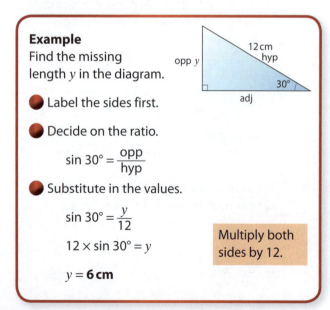

🔴 Label the sides first.

🔴 Decide on the ratio.

$$\sin 30° = \frac{\text{opp}}{\text{hyp}}$$

🔴 Substitute in the values.

$$\sin 30° = \frac{y}{12}$$

$$12 \times \sin 30° = y$$

Multiply both sides by 12.

$$y = \textbf{6 cm}$$

Finding an Angle

Example
Calculate angle ABC.

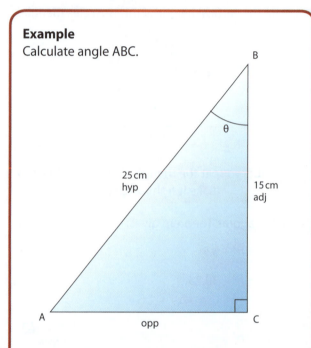

Label the sides and decide on the ratio.

$$\cos \theta = \frac{\text{adj}}{\text{hyp}}$$

$$\cos \theta = \frac{15}{25}$$

$$\cos \theta = 0.6$$

$$\theta = \cos^{-1} 0.6$$

To find the angle, you usually use the second function on your calculator.

$$= \textbf{53.13°} \text{ (2dp)}$$

It is important you know how to use your calculator when working out trigonometry questions. Also, do not round off before the end of a question.

Angle of Elevation

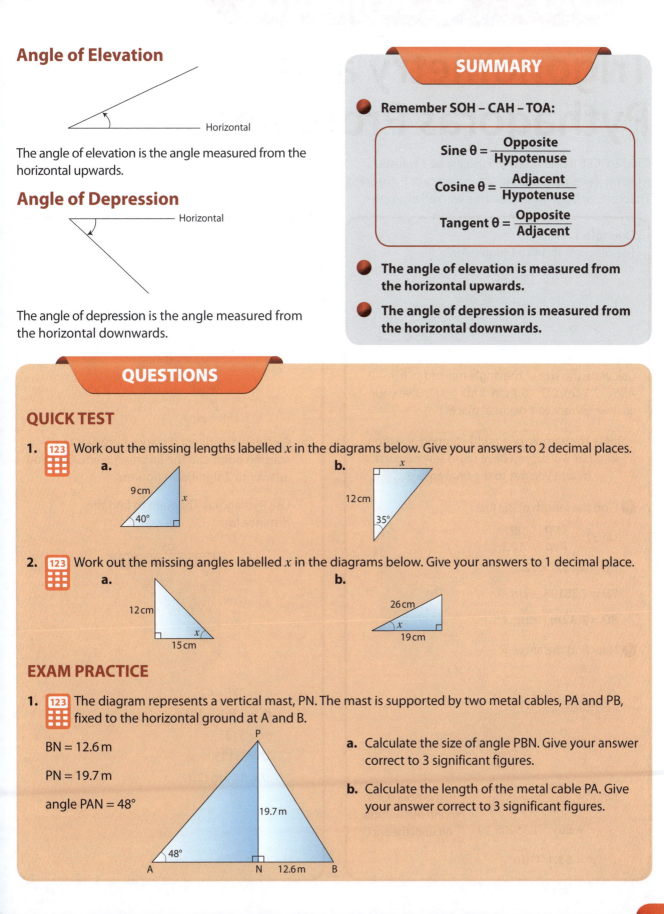

The angle of elevation is the angle measured from the horizontal upwards.

Angle of Depression

The angle of depression is the angle measured from the horizontal downwards.

SUMMARY

● Remember **SOH – CAH – TOA:**

> **Sine** $\theta = \dfrac{\textbf{Opposite}}{\textbf{Hypotenuse}}$
>
> **Cosine** $\theta = \dfrac{\textbf{Adjacent}}{\textbf{Hypotenuse}}$
>
> **Tangent** $\theta = \dfrac{\textbf{Opposite}}{\textbf{Adjacent}}$

● **The angle of elevation is measured from the horizontal upwards.**

● **The angle of depression is measured from the horizontal downwards.**

QUESTIONS

QUICK TEST

1. Work out the missing lengths labelled x in the diagrams below. Give your answers to 2 decimal places.

   a. 9 cm, 40°, x

   b. x, 12 cm, 35°

2. Work out the missing angles labelled x in the diagrams below. Give your answers to 1 decimal place.

   a. 12 cm, 15 cm, x

   b. 26 cm, x, 19 cm

EXAM PRACTICE

1. The diagram represents a vertical mast, PN. The mast is supported by two metal cables, PA and PB, fixed to the horizontal ground at A and B.

   BN = 12.6 m

   PN = 19.7 m

   angle PAN = 48°

   P, 19.7 m, 48°, A, N, 12.6 m, B

   a. Calculate the size of angle PBN. Give your answer correct to 3 significant figures.

   b. Calculate the length of the metal cable PA. Give your answer correct to 3 significant figures.

Trigonometry and Pythagoras Problems

On the GCSE paper there will usually be a question which involves you applying trigonometry or Pythagoras' Theorem, or both. Here are some worked examples.

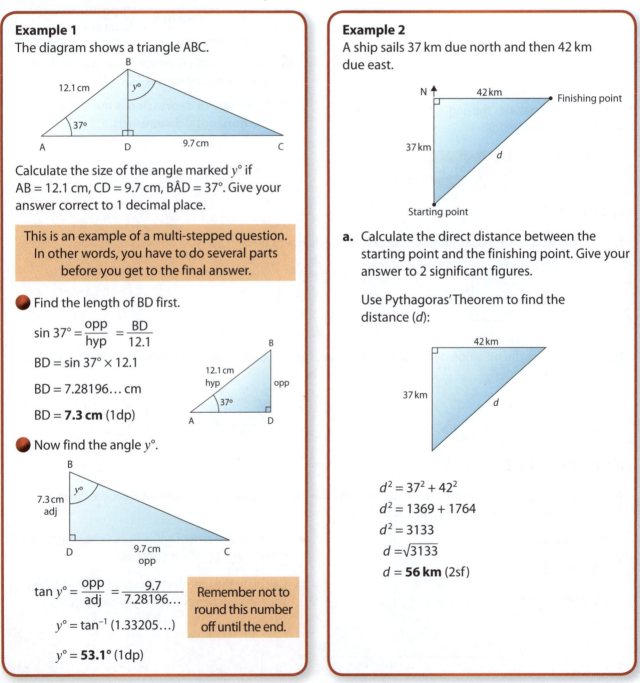

Example 1

The diagram shows a triangle ABC.

Calculate the size of the angle marked $y°$ if AB = 12.1 cm, CD = 9.7 cm, BÂD = 37°. Give your answer correct to 1 decimal place.

> This is an example of a multi-stepped question. In other words, you have to do several parts before you get to the final answer.

● Find the length of BD first.

$$\sin 37° = \frac{\text{opp}}{\text{hyp}} = \frac{BD}{12.1}$$

$BD = \sin 37° \times 12.1$

$BD = 7.28196\ldots$ cm

$BD = \textbf{7.3 cm}$ (1dp)

● Now find the angle $y°$.

$$\tan y° = \frac{\text{opp}}{\text{adj}} = \frac{9.7}{7.28196\ldots}$$

> Remember not to round this number off until the end.

$y° = \tan^{-1}(1.33205\ldots)$

$y° = \textbf{53.1°}$ (1dp)

Example 2

A ship sails 37 km due north and then 42 km due east.

a. Calculate the direct distance between the starting point and the finishing point. Give your answer to 2 significant figures.

Use Pythagoras' Theorem to find the distance (d):

$d^2 = 37^2 + 42^2$

$d^2 = 1369 + 1764$

$d^2 = 3133$

$d = \sqrt{3133}$

$d = \textbf{56 km}$ (2sf)

Example 2 (cont.)

b. Calculate the bearing of the ship from its starting point. Give your answer to the nearest degree.

Use trigonometry to find the bearing:

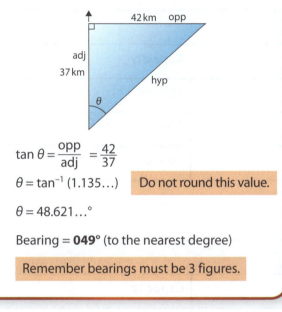

$\tan \theta = \dfrac{\text{opp}}{\text{adj}} = \dfrac{42}{37}$

$\theta = \tan^{-1}(1.135\ldots)$ Do not round this value.

$\theta = 48.621\ldots°$

Bearing = **049°** (to the nearest degree)

Remember bearings must be 3 figures.

SUMMARY

- **Pythagoras' Theorem:** $a^2 + b^2 = c^2$
- **The three trigonometric ratios are:**

  $\text{Sine } \theta = \dfrac{\textbf{Opposite}}{\textbf{Hypotenuse}}$

  $\text{Cosine } \theta = \dfrac{\textbf{Adjacent}}{\textbf{Hypotenuse}}$

  $\text{Tangent } \theta = \dfrac{\textbf{Opposite}}{\textbf{Adjacent}}$

- **Bearings are measured from the North in a clockwise direction; they are written using 3 figures.**

QUESTIONS

QUICK TEST

1. [123] Work out the area of the triangle. Give your answer to 3 significant figures.

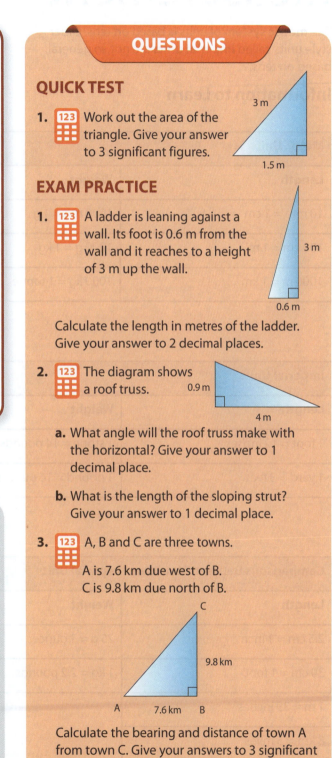

EXAM PRACTICE

1. [123] A ladder is leaning against a wall. Its foot is 0.6 m from the wall and it reaches to a height of 3 m up the wall.

Calculate the length in metres of the ladder. Give your answer to 2 decimal places.

2. [123] The diagram shows a roof truss.

a. What angle will the roof truss make with the horizontal? Give your answer to 1 decimal place.

b. What is the length of the sloping strut? Give your answer to 1 decimal place.

3. [123] A, B and C are three towns.

A is 7.6 km due west of B.
C is 9.8 km due north of B.

Calculate the bearing and distance of town A from town C. Give your answers to 3 significant figures.

Measurement

The **metric system** of units is based on tens. Older style units called **imperial units** are not, in general, based on tens.

Information to Learn

Metric Units		
Length	**Weight**	**Capacity**
10 mm = 1 cm	1000 mg = 1 g	1000 ml = 1 l
100 cm = 1 m	1000 g = 1 kg	100 cl = 1 l
1000 m = 1 km	1000 kg = 1 tonne	1000 cm^3 = 1 l

Imperial Units		
Length	**Weight**	**Capacity**
1 foot = 12 inches	1 stone = 14 pounds (lb)	20 fluid oz = 1 pint
1 yard = 3 feet	1 pound = 16 ounces (oz)	8 pints = 1 gallon

Comparisons Between Metric and Imperial Units		
Length	**Weight**	**Capacity**
2.5 cm ≈ 1 inch	25 g ≈ 1 ounce	1 litre ≈ $1\frac{3}{4}$ pints
30 cm ≈ 1 foot	1 kg ≈ 2.2 pounds	4.5 litres ≈ 1 gallon
1 m ≈ 39 inches		
8 km ≈ 5 miles		
N.B. These comparisons are only approximate.		

Compound Measures

Compound measures involve a combination of basic measures.

Speed

Units of speed are:

- metres per second (m/s)
- kilometres per hour (km/h)
- miles per hour (mph)

$$\text{Speed } (s) = \frac{\text{distance travelled } (d)}{\text{time taken } (t)}$$

Rearranging gives:

$$\text{Time taken} = \frac{\text{distance travelled}}{\text{speed}}$$

$$\text{Distance travelled} = \text{speed} \times \text{time taken}$$

Remember to check the units before starting a question. Change them if necessary.

Examples

1. A car travels 80 km in 1 hour 20 minutes. Find the speed in km/h.

$$s = \frac{d}{t}$$

$$s = \frac{80}{1.\dot{3}}$$

$$s = \textbf{60 km/h}$$

> Change the time into hours. 20 minutes is $\frac{20}{60}$ of 1 hour.

2. Miss Fitzgerald drives 40 miles to work. On Wednesday her journey to work took 50 minutes. On Thursday the average speed of her journey to work was 54 km/h.

   Did Miss Fitzgerald drive more quickly to work on Wednesday or Thursday?

   Speed on Wednesday is $\frac{40}{\frac{50}{60}} = 48$ mph

   Since 1 km = $\frac{5}{8}$ mile

   Speed on Thursday is $\frac{5}{8} \times 54 = 33.75$ mph

   Miss Fitzgerald drove more quickly to work on Wednesday.

Density

$$\text{Density} = \frac{\text{mass}}{\text{volume}}, \quad \text{Volume} = \frac{\text{mass}}{\text{density}},$$

$$\text{Mass} = \text{volume} \times \text{density}$$

SUMMARY

- Compound measures involve a combination of basic measures.
- Units of speed are metres per second (m/s), kilometres per hour (km/h), miles per hour (mph).
- Speed $= \dfrac{\text{distance travelled}}{\text{time taken}}$
- Density $= \dfrac{\text{mass}}{\text{volume}}$

QUESTIONS

QUICK TEST

1. Change 3200 g into kilograms.
2. Change 4 kg into pounds.
3. Change 6 litres into pints.
4. [123] Find the time taken for a car to travel 240 miles at the average speed of 70 mph.

EXAM PRACTICE

1. [123] Two solids each have a volume of 2.5 m³. The density of solid A is 320 kg per m³. The density of solid B is 288 kg per m³.

   Calculate the difference in the masses of the solids.

2. [123] Josie is driving in her car at a speed of 70 mph. Jack is driving in his car at a speed of 120 km/h.

   Who is travelling at the fastest speed? You must show full working out to explain your answer.

Areas of Plane Shapes 1

The area of a 2D shape is the amount of flat space that it covers. Common units of area are mm^2, cm^2, m^2.

Key Formulae

Area of a Rectangle

Area = length × width

$A = l \times w$

Area of a Parallelogram

Area = base × perpendicular height

$A = b \times h$

Remember to use the perpendicular height, not the slant height.

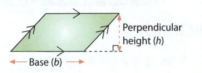

Area of a Triangle

Area = $\frac{1}{2}$ base × perpendicular height

$A = \frac{1}{2} \times b \times h$

Area of a Trapezium

Area = $\frac{1}{2}$ × (sum of parallel sides) × perpendicular height between them

$A = \frac{1}{2} \times (a + b) \times h$

$A = \frac{1}{2} (a + b) h$

> You don't need to remember this one as it's on the formula sheet.

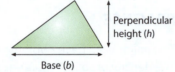

Example

The diagram shows a trapezium made of metal.

A rectangular piece of metal is cut out of the trapezium as shown.

Work out the area of the remaining metal, shown shaded in the diagram below:

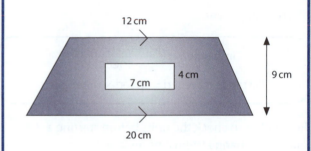

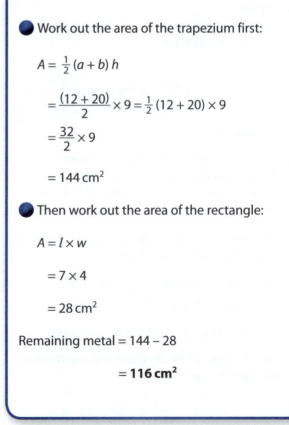

● Work out the area of the trapezium first:

$A = \frac{1}{2} (a + b) h$

$= \frac{(12 + 20)}{2} \times 9 = \frac{1}{2} (12 + 20) \times 9$

$= \frac{32}{2} \times 9$

$= 144 \ cm^2$

● Then work out the area of the rectangle:

$A = l \times w$

$= 7 \times 4$

$= 28 \ cm^2$

Remaining metal = 144 − 28

$= \mathbf{116 \ cm^2}$

More Complex Shapes

Work out the area of more complex shapes by splitting them into simple shapes.

Example
The diagram below shows the plan of a garden.

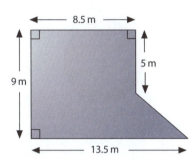

Lawn seed is sown to cover the garden. Lawn seed comes in 500 g packets and covers 15 m². A packet of lawn seed costs £6.25. Work out the total cost of the lawn seed needed.

Work out the area of the garden by splitting it into a rectangle and triangle.

Area of rectangle = $l \times w$

$$= 9 \times 8.5$$
$$= 76.5 \, m^2$$

Area of triangle = $\frac{1}{2} \times b \times h$

$$= \frac{1}{2} \times 5 \times 4$$

> Base of triangle
> $= 13.5 - 8.5 = 5$ m

$$= 10 \, m^2$$

Total area = 76.5 + 10
$$= 86.5 \, m^2$$

Number of packets of lawn seed needed:
$\frac{86.5}{15} = 5.76$ packets, hence 6 packets of lawn seed are needed.

Cost = £6.25 × 6
$$= £37.50$$

SUMMARY

- Area is measured in square units.
- Area of a rectangle: $A = l \times w$
  Area of a parallelogram: $A = b \times h$
  Area of a triangle: $A = \frac{1}{2} \times b \times h$
  Area of a trapezium: $A = \frac{1}{2}(a + b)\,h$
- Work out the area of more complex shapes by splitting them into simple shapes.

QUESTIONS

QUICK TEST

1. Find the areas of these shapes:

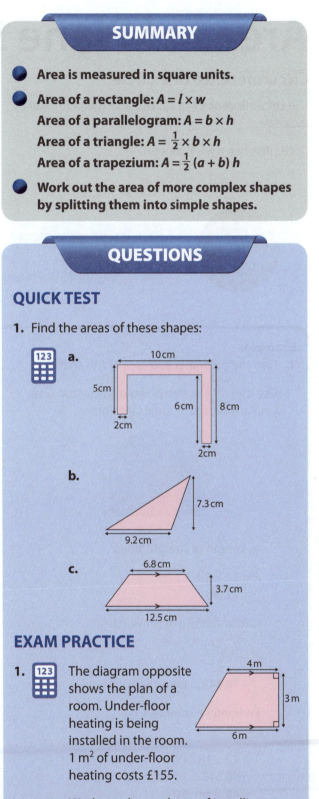

a.

b.

c.

EXAM PRACTICE

1. The diagram opposite shows the plan of a room. Under-floor heating is being installed in the room. 1 m² of under-floor heating costs £155.

Work out the total cost of installing under-floor heating for the whole room.

Areas of Plane Shapes 2

Circumference of a Circle

The **circumference** of a circle is also the perimeter of the circle.

Circumference = π × diameter

= 2 × π × radius

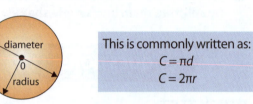

> This is commonly written as:
> $C = \pi d$
> $C = 2\pi r$

Area of a Circle

Area = π × (radius)²

$A = \pi \times r^2$

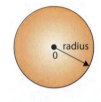

Examples

1. Find the perimeter and area of this shape.

Use the 🔲 π button on your calculator. Give your answers to 3 significant figures.

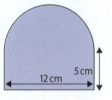

5 cm
12 cm

a. Length of straight sides:

$P = 12 + 5 + 5$

$\quad = 22$ cm

Circumference of semicircle:

$C = \dfrac{\pi \times 12}{2}$

$\quad = 18.849\ldots$ cm

Perimeter of shape:

$= P + C$

$= 22 + 18.849\ldots$

$= 40.849\ldots$

$= \mathbf{40.8\ cm}$ (3sf)

b. Area of rectangle $= l \times w$

$\quad = 12 \times 5$

$\quad = 60\ cm^2$

Area of semicircle $= \dfrac{\pi \times r^2}{2}$

$= \dfrac{\pi \times 6^2}{2}$

$= 56.548\ldots cm^2$

Total area $= 60 + 56.548$

$= 116.548\ldots$

$= \mathbf{117\ cm^2}$ (3sf)

2. A circle has an area of 60 cm². Find the radius of the circle, giving your answer to 3 significant figures. Use π = 3.142

$A = \pi \times r^2$

$60 = 3.142 \times r^2$

> Substitute the values into the formula.

$\dfrac{60}{3.142} = r^2$

> Divide both sides by 3.142

$r^2 = 19.0961\ldots$

$r = \sqrt{19.0961\ldots}$

> Take the square root to find r.

$r = \mathbf{4.37\ cm}$ (3sf)

Examples (cont.)

3. Work out the area of the dark shaded region.

Use the $\boxed{\pi}$ button on your calculator.

Give your answer to 3 significant figures.

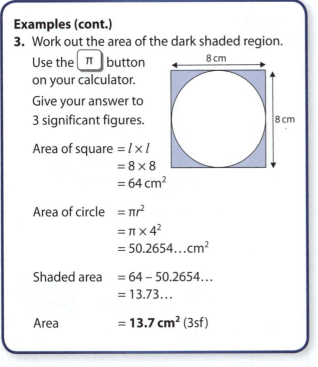

Area of square $= l \times l$
$\qquad = 8 \times 8$
$\qquad = 64 \text{ cm}^2$

Area of circle $= \pi r^2$
$\qquad = \pi \times 4^2$
$\qquad = 50.2654\ldots \text{cm}^2$

Shaded area $= 64 - 50.2654\ldots$
$\qquad = 13.73\ldots$

Area $\qquad = \textbf{13.7 cm}^2$ (3sf)

Surface Area of a Cylinder

The total **surface area** of a cylinder can be found by drawing a net.

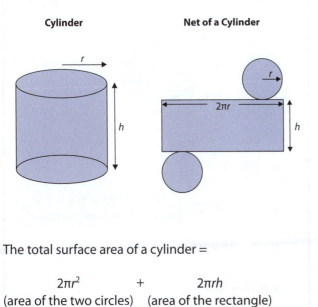

Cylinder   **Net of a Cylinder**

The total surface area of a cylinder =

$\qquad 2\pi r^2 \qquad + \qquad 2\pi rh$
(area of the two circles) (area of the rectangle)

QUESTIONS

QUICK TEST

1. Find the circumference and area of these circles. Use $\pi = 3.142$

   $\boxed{123}$ Give your answers to 2 decimal places.

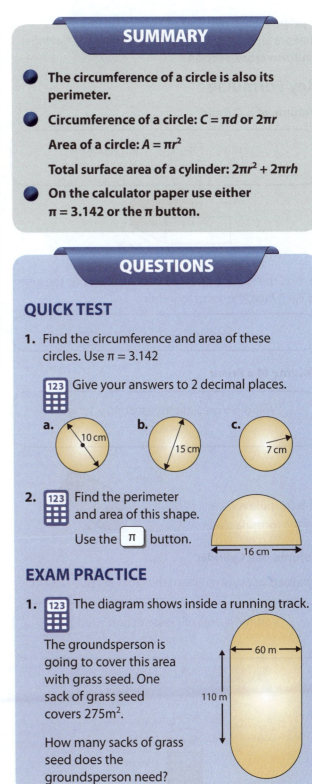

   a. 10 cm   b. 15 cm   c. 7 cm

2. $\boxed{123}$ Find the perimeter and area of this shape.

   Use the $\boxed{\pi}$ button.

   16 cm

EXAM PRACTICE

1. $\boxed{123}$ The diagram shows inside a running track.

   The groundsperson is going to cover this area with grass seed. One sack of grass seed covers 275m².

   60 m

   110 m

   How many sacks of grass seed does the groundsperson need?

Volumes of Prisms

A **prism** is any solid that can be cut up into slices that are all the same shape. This is known as having a **uniform cross-section**.

Key Formulae

Volume of a Cuboid

> Volume = length × width × height
> $V = l \times w \times h$

To find the surface area of a cuboid, work out the area of each face, then add together.

> Surface area = $2hl + 2hw + 2lw$

Volume of a Prism

> Volume = area of cross-section × length
> $V = A \times l$

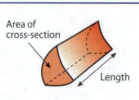

This formula is given on the formula sheet.

Volume of a Cylinder

Cylinders are prisms where the cross-section is a circle.

> Volume = area of cross-section × length
> $V = \pi r^2 \times h$

Area of circle   Height or length

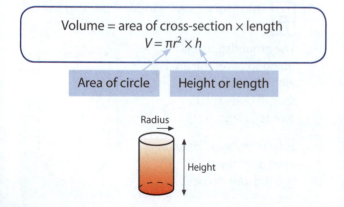

Examples

1. Find the volume of this prism.

$V = A \times l$

$V = (\frac{1}{2} \times b \times h) \times l$ The area of the cross-section is the area of a triangle.

$V = (\frac{1}{2} \times 7.5 \times 12) \times 15$

$V = \textbf{675 cm}^3$ Cubic units for volume

2. Find the volume of this cylinder. Leave your answer in terms of π.

$V = \pi^2 r \times h$

$V = \pi \times 6^2 \times 10$ Remember to halve the diameter to find the radius.

$V = \pi \times 36 \times 10$

$V = \textbf{360}\boldsymbol{\pi}\ \textbf{cm}^3$

Solving More Difficult Problems

Sometimes you will be given a more difficult problem to solve.

Example

A gold bar has a cross-section in the shape of a trapezium. The length of the bar is 15 cm.

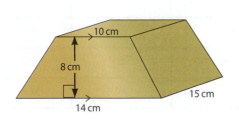

The prism is made out of gold. Gold has a density of 19.3 g/cm³.

Work out the mass of the prism.

Volume of prism = area of cross-section × length

$$V = (\tfrac{1}{2}(a+b)h) \times l$$

> Work out the volume first.

$$= \tfrac{1}{2} \times (10 + 14) \times 8 \times 15$$

$$= 1440 \text{ cm}^3$$

$$\text{Density} = \frac{\text{mass}}{\text{volume}}$$

> Rearrange the formula for density to find the mass.

Mass = volume × density

Mass = 1440 × 19.3

$$= 27\,792 \text{ grams}$$

$$= \textbf{27.792 kg}$$

SUMMARY

- **Volume is measured in cubic units.**
- **A prism has a uniform cross-section.**
- **Volume of a cuboid:** $V = l \times w \times h$

  Volume of a prism: $V = A \times l$

  Volume of a cylinder: $V = \pi r^2 \times h$

QUESTIONS

QUICK TEST

1. Work out the volumes of these prisms. Give your answers to 1dp. Use π = 3.142

   a.

   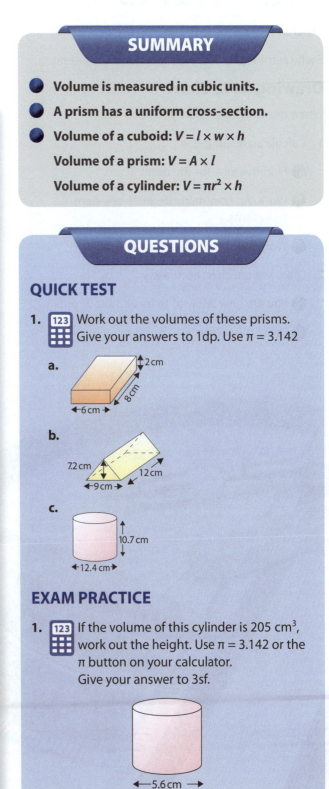

   2 cm, 8 cm, 6 cm

   b.

   7.2 cm, 12 cm, 9 cm

   c.

   10.7 cm, 12.4 cm

EXAM PRACTICE

1. If the volume of this cylinder is 205 cm³, work out the height. Use π = 3.142 or the π button on your calculator. Give your answer to 3sf.

   5.6 cm

Pie Charts

Pie charts are circles split up into sectors.

Each sector represents a certain number of items.

Drawing a Pie Chart

When drawing a pie chart you need to:

1. Calculate the angles:

    Find the total for the items listed.

    Work out how many degrees one item represents.

    Work out the degrees for each category.

2. Draw the pie chart accurately:

    You are only allowed to be at most 2° out!

    Label the sectors.

Example

The favourite subjects of 24 students are shown in the table. Draw a pie chart for this information.

Subject	Frequency	Angle
Maths	9	135°
English	4	60°
Art	5	75°
Geography	6	90°
	24	**360°**

24 students = 360°

$$1 \text{ student} = \frac{360°}{24} = 15°$$

Multiply each frequency by 15° to find the angles. Now draw the pie chart carefully.

The angles must be accurate and each sector must be labelled.

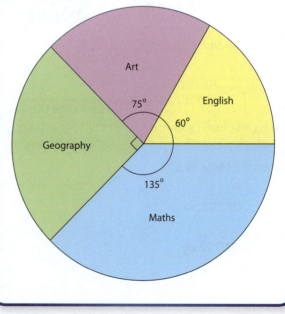

Interpreting a Pie Chart

When interpreting a pie chart you need to measure the angles carefully if they are not given.

Example

The pie chart shows the favourite sports of 18 students.

How many students like:

a. tennis?

b. football?

c. hockey?

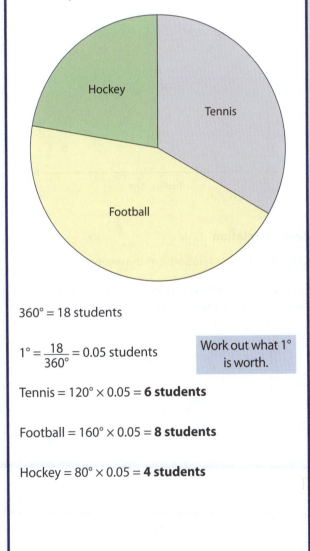

360° = 18 students

$1° = \dfrac{18}{360°} = 0.05$ students

> Work out what 1° is worth.

Tennis = 120° × 0.05 = **6 students**

Football = 160° × 0.05 = **8 students**

Hockey = 80° × 0.05 = **4 students**

SUMMARY

- When drawing a pie chart, you need to work out how many degrees one item represents and then multiply by the number of items in each category to find the angles.

- Always draw the pie chart accurately and label the sectors.

- When interpreting a pie chart, you need to work out what 1° is worth and then multiply by the total degrees in each angle to find the number of items in each category.

QUESTIONS

QUICK TEST

1. Draw a pie chart for this set of data.

Favourite Colour	Frequency
Blue	15
Red	9
Black	5
Green	7

EXAM PRACTICE

1. 🔢 The local authority wish to build a new school and a new retirement home. They look at the age profile of two towns.

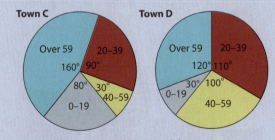

Town C has 75 40–59 year olds.
Town D has 500 40–59 year olds.

Using the information in the pie charts, state with reasons and calculations which town should have the school and which town should have the retirement home.

Scatter Diagrams and Correlation

Scatter diagrams are used to show two sets of data at the same time. They are important because they show the **correlation** (connection) between the sets.

Types of Correlation

Positive Correlation

Both variables are increasing.

For example, the taller you are, the more you probably weigh.

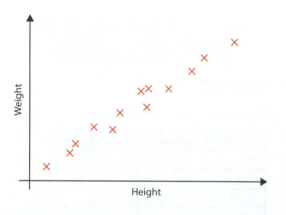

Negative Correlation

As one variable increases, the other decreases.

For example, as the temperature increases, the sale of woollen hats probably decreases.

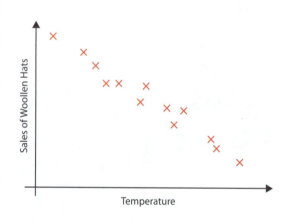

Zero Correlation

Little or no linear relationship between the variables.

For example, there is no connection between your height and your mathematical ability.

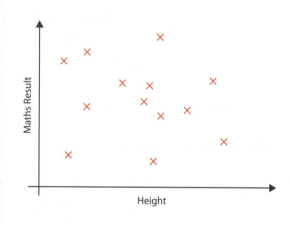

Line of Best Fit

The **line of best fit** goes as close as possible to all the points.

There is roughly an equal number of points above the line and below it.

The scatter diagram below shows the Science and Maths percentages scored by some students.

- The line of best fit goes in the direction of the data.
- The line of best fit can be used to estimate results.
- We can estimate that a student with a Science percentage of 30 would get a Maths percentage of about 17.
- We can estimate that a student with a Maths percentage of 50 would get a Science percentage of about 54.

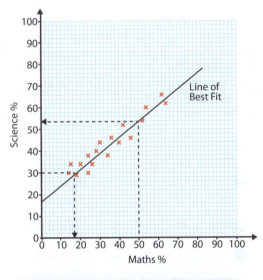

SUMMARY

- **Positive correlation: both variables are increasing.**
- **Negative correlation: as one variable increases, the other decreases.**
- **Zero correlation: little or no linear correlation between variables.**
- **A line of best fit should be as close as possible to all the points. It is in the direction of the data.**

QUESTIONS

QUICK TEST

Decide whether these statements are **true** or **false**.

1. There is a positive correlation between the weight of a book and the number of pages.

2. There is no correlation between the height you climb up a mountain and the temperature.

3. There is a negative correlation between the age of a used car and its value.

4. There is a positive correlation between the height of some students and the size of their feet.

5. There is no correlation between the weight of some students and their History GCSE results.

EXAM PRACTICE

1. The scatter diagram shows the age of some cars and their values.

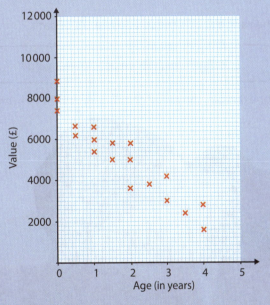

a. Draw a line of best fit on the diagram.

b. Use your line of best fit to estimate the age of a car when its value is £5000.

c. Use your line of best fit to estimate the value of a $3\frac{1}{2}$ year-old car.

Averages 1

Averages of Discrete Data

Averages are used to give an idea of a 'typical' value for a set of data. **Discrete data** has an exact value.

You should know these types of averages:

1. **Mean** – the most commonly used average:

> $$\text{Mean} = \frac{\text{sum of a set of values}}{\text{the number of values used}}$$

For example, the mean of 1, 2, 3, 3, 1 is

$$\frac{1 + 2 + 3 + 3 + 1}{5} = 2$$

2. **Median** – the middle value when the values are put in order of size.

For example, the median of 2, 2, 3, 3, 7, 9, 11 is 3

3. **Mode** – the one that occurs the most often.

For example, the mode of 2, 2, 2, 3, 5, 7 is 2

The **range** shows the spread of a set of data.

> $$\text{Range} = \text{highest value} - \text{lowest value}$$

For example, the range of 1, 2, 3, 4, 7, 10 is $10 - 1 = 9$

Finding Averages from a Frequency Table

When the information is in a frequency table, finding the averages is a little more difficult.

To find the mean of a frequency table, we use:

> $$\text{mean } (\bar{x}) = \frac{\Sigma fx}{\Sigma f}$$
>
> Σ means the sum of
> f represents the frequency
> $\bar{x}$ represents the mean

Example

The table shows the shoe sizes of a group of students. Find the mean, range, mode and median.

Shoe Size (x)	3	4	5	6	7
Frequency (f)	5	18	22	15	5

Mean shoe size $\text{Mean} = (\bar{x}) = \dfrac{\Sigma fx}{\Sigma f}$

🟢 Multiply the frequency, f, by x
$= (3 \times 5) + (4 \times 18) + (5 \times 22) + (6 \times 15) + (7 \times 5)$

🟢 Add up the frequency $= 5 + 18 + 22 + 15 + 5$

🟢 Divide the sum of fx by the sum of f:

$$\frac{(3 \times 5) + (4 \times 18) + (5 \times 22) + (6 \times 15) + (7 \times 5)}{5 + 18 + 22 + 15 + 5}$$

$$= \frac{322}{65} \quad = \textbf{4.95} \text{ (2dp)}$$

Range   Range = highest value – lowest value

Range $= 7 - 3 = \textbf{4}$

Mode   Modal shoe size = **5** (highest frequency)

Median   Median shoe size: $\frac{\Sigma f + 1}{2}$ tells you how many places along the list to go.

33rd person has size 5, so median = **5**

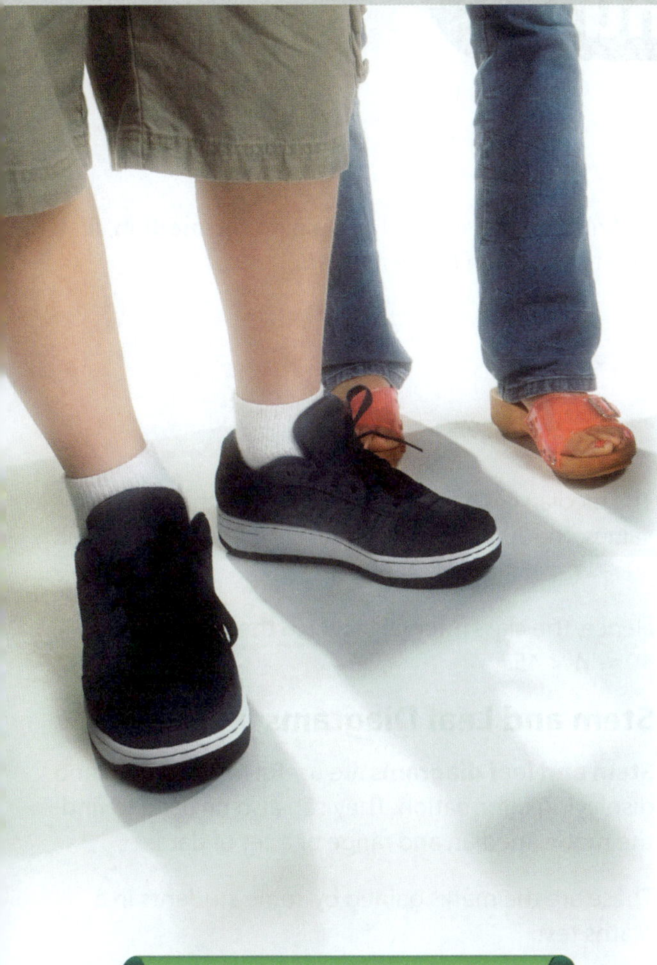

- For discrete data:
    - Mean = $\dfrac{\text{sum of a set of values}}{\text{the number of values used}}$
    - Median = middle value when values are put in order of size
    - Mode = the value that occurs most often
- Range = highest value – lowest value
- Find the mean from a frequency table:

$$\text{mean } (\bar{x}) = \frac{\Sigma fx}{\Sigma f}$$

Σ means the sum of
f represents the frequency
$\bar{x}$ represents the mean

QUESTIONS

QUICK TEST

1. 123 Find the mean and range of this data:

      a. 2, 7, 9, 3, 6, 4, 5, 2   **b.** 7, 9, 11, 15, 2, 1, 6, 12, 19, 13

2. Find the median and mode of this data:

      a. 2, 7, 1, 4, 2, 2, 3, 7, 9, 2   **b.** 6, 9, 11, 11, 13, 6, 9, 6, 6, 4, 1

EXAM PRACTICE

1. Reece made up this table for the number of minutes (to the nearest minute) it took some students to complete a Maths problem.

123 Calculate:
   a. the mean
   b. the median
   c. the mode
   d. the range

Number of Minutes to Solve Problem	5	6	7	8	9	10
Frequency	4	7	10	4	3	1

Averages 2

Averages of Continuous Data

When the data is grouped into **class intervals**, the exact data is not known.

We estimate the **mean** by using the midpoints of the class intervals.

> Add in 2 extra columns – one for the midpoint and one for fx.

Weight (W kg)	Frequency (f)	Midpoint (x)	fx
$30 \leqslant W < 35$	6	32.5	195
$35 \leqslant W < 40$	14	37.5	525
$40 \leqslant W < 45$	22	42.5	935
$45 \leqslant W < 50$	18	47.5	855
	60		**2510**

Σf Σfx

For continuous data:

> $$\bar{x} = \frac{\Sigma fx}{\Sigma f}$$
>
> Σ means the sum of
> f represents the frequency
> $\bar{x}$ represents the mean
> x represents the midpoint of the class interval

$$\bar{x} = \frac{\Sigma fx}{\Sigma f}$$

$$\bar{x} = \frac{2510}{60}$$

$$\bar{x} = 41.8\dot{3} \text{ (2dp)}$$

Modal class is $40 \leqslant W < 45$

This class interval has the highest frequency.

To find the class interval containing the **median**, first find the position of the median.

$$= \frac{\Sigma f + 1}{2}$$

$$= \frac{60 + 1}{2} = 30.5$$

The median lies half way between the 30th and 31st values. The 30th and 31st values are in the class interval: $40 \leqslant W < 45$.

Hence, the class interval in which the median lies is $40 \leqslant W < 45$.

Stem and Leaf Diagrams

Stem and leaf diagrams are useful for recording and displaying information. They can also be used to find the mode, median and range of a set of data.

These are the marks gained by some students in a Maths test:

52	45	63	67
75	57	68	67
60	59	67	

In an ordered stem and leaf diagram it would look like this:

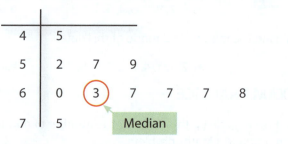

Key $6 \mid 3 = 63$ marks

The median is the sixth value = 63 marks

The mode is 67 marks

The range is $75 - 45 = 30$ marks

If the same students sat a second Maths test, their results could be put into a **back-to-back stem and leaf diagram**. These are very useful when comparing two sets of data.

			Test 2		Test 1				
	9	6	2	4	5				
9	7	5	1	1	5	2	7	9	
		3	1	0	6	0	3	7	7 7 8
				7	5				

Key   Test 1   5|2 = 52 marks
        Test 2   1|5 = 51 marks

Comparing the data in the back-to-back stem and leaf, we can say that:

'In test 2, the median score of 55 marks is lower than the median score in test 1 of 63 marks. The range of the scores in test 2 is 21 marks, which is lower than the range of the scores in test 1 of 30 marks. So on average, the students did better in test 1 but their scores were more variable than in test 2.'

QUESTIONS

QUICK TEST

1. 123 The heights, *h* cm, of some students are shown in the table.

Height (*h* cm)	Frequency	Midpoint	*fx*
$140 \leqslant h < 145$	4		
$145 \leqslant h < 150$	9		
$150 \leqslant h < 155$	15		
$155 \leqslant h < 160$	6		

Calculate an estimate for the mean of this data.

2. **a.** Draw an accurate stem and leaf diagram of this data.

27	28	36	42	50	18
25	31	39	25	49	31
33	27	37	25	47	40
7	31	26	36	9	42

   b. What is the median of this data?

EXAM PRACTICE

1. The table shows information about the number of hours that 50 children watched television for last week.

Work out an estimate for the mean number of hours the children watched television.

Number of Hours (*h*)	Frequency
$0 \leqslant h < 2$	3
$2 \leqslant h < 4$	6
$4 \leqslant h < 6$	22
$6 \leqslant h < 8$	13
$8 \leqslant h < 10$	6

Cumulative Frequency Graphs

With a **cumulative frequency graph** it is possible to estimate the median of grouped data and the interquartile range.

Example

The cumulative frequency table opposite shows the marks of 94 students in a Maths exam.

a. Complete the cumulative frequency table for this data.

b. Draw a cumulative frequency graph for this data.

To do this we must plot the upper boundary of each class interval on the x-axis and the cumulative frequency on the y-axis.
Plot (20, 2) (30, 8) (40, 18) …
Join the points with a smooth curve.
Since no students had less than zero marks, the graph starts at (0, 0).

c. Use the cumulative frequency graph to find the median and interquartile range of the data.

> Cumulative frequency is a running total of all the frequencies.

Mark (m)	Frequency	Mark	Cumulative Frequency
$0 < m \leqslant 20$	2	$\leqslant 20$	2 (2)
$20 < m \leqslant 30$	6	$\leqslant 30$	8 (2 + 6)
$30 < m \leqslant 40$	10	$\leqslant 40$	18 (8 + 10)
$40 < m \leqslant 50$	17	$\leqslant 50$	35 (18 + 17)
$50 < m \leqslant 60$	24	$\leqslant 60$	59 (35 + 24)
$60 < m \leqslant 70$	17	$\leqslant 70$	76 (59 + 17)
$70 < m \leqslant 80$	11	$\leqslant 80$	87 (76 + 11)
$80 < m \leqslant 90$	4	$\leqslant 90$	91 (87 + 4)
$90 < m \leqslant 100$	3	$\leqslant 100$	94 (91 + 3)

> This means that 94 students had a score of 100 or less.

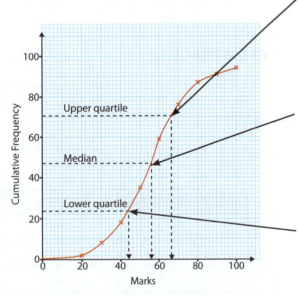

The **upper quartile** is three quarters of the way into the distribution:
$$\frac{3}{4} \times 94 = 70.5$$
Read across from 70.5 and down to the horizontal axis.
Upper quartile ≈ 67 marks

The **median** splits the data into two halves – the lower 50% and the upper 50%.
Median $= \frac{1}{2} \times$ cumulative frequency
$$= \frac{1}{2} \times 94 = 47.$$
Read across from 47. Median ≈ 56 marks

The **lower quartile** is the value one quarter of the way into the distribution:
$$\frac{1}{4} \times 94 = 23.5$$
Read across from 23.5   Lower quartile ≈ 44 marks

Interquartile range

= upper quartile − lower quartile
= 67 − 44 = 23 marks

A large interquartile range indicates that the 'middle half' of the data is widely spread about the median.

A small interquartile range indicates that the 'middle half' of the data is concentrated about the median.

Box Plots

● Box plots are sometimes known as box and whisker diagrams.

● Cumulative frequency graphs are not easy to compare; a box plot shows the interquartile range as a box, and the highest and lowest values as whiskers. Comparing the spread of data is then easier.

Example
The box plot of the cumulative frequency graph opposite would look like this:

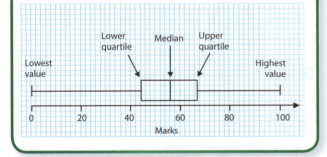

QUICK TEST

1. The box plot below shows the times in minutes to finish an assault course.

Use the box plot to complete the table below.

	Time in Minutes
Median time	
Lower quartile	
Interquartile range	
Longest time	

EXAM PRACTICE

1. Students in 9A and 9B took the same test. Their results were used to draw the following box plots.

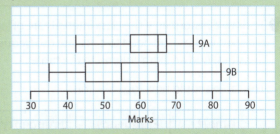

a. In which class was the student who scored the highest mark?

b. In which class did the students perform better in the test? You must give a reason for your answer.

SUMMARY

● **To draw a cumulative frequency graph, plot the upper boundary of each class interval on the *x*-axis and the cumulative frequency on the *y*-axis.**

● **Interquartile range = upper quartile – lower quartile**

● **Use box plots to compare data.**

Histograms

In a **histogram** the area of a bar is proportional to the frequency.

● If the class intervals have **equal widths**, frequency can be used for the height of the bar.

● If the class intervals have **unequal widths**, the height of the bar is adjusted by using **frequency density** and the area of the bar is equal to the frequency.

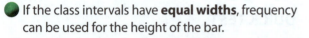

$$\text{Frequency density} = \frac{\text{frequency}}{\text{class width}}$$

The **modal class** is the class interval with the highest bar.

Drawing Histograms

The table below shows the time in seconds it takes people to swim 100 metres.

Time, t (seconds)	Frequency
$100 < t \leqslant 110$	2
$110 < t \leqslant 140$	24
$140 < t \leqslant 160$	42
$160 < t \leqslant 200$	50
$200 < t \leqslant 220$	24
$220 < t \leqslant 300$	20

● To draw a histogram if the class intervals are of different widths, you need to calculate the frequency densities. Add an extra column to the table.

The table should now look like this:

Time, t (seconds)	Frequency	Frequency Density
$100 < t \leqslant 110$	2	$2 \div 10 = 0.2$
$110 < t \leqslant 140$	24	$24 \div 30 = 0.8$
$140 < t \leqslant 160$	42	$42 \div 20 = 2.1$
$160 < t \leqslant 200$	50	$50 \div 40 = 1.25$
$200 < t \leqslant 220$	24	$24 \div 20 = 1.2$
$220 < t \leqslant 300$	20	$20 \div 80 = 0.25$

● Draw on graph paper – make sure there are no gaps between bars.

The histogram should look like this:

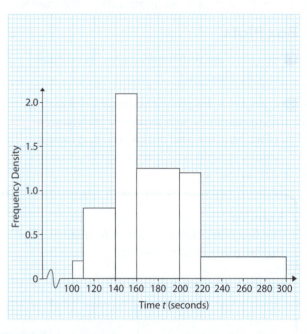

● Sometimes you will be asked to read from a histogram. In this case, rearrange the formula for frequency density.

$$\text{Frequency} = \text{frequency density} \times \text{class width}$$

SUMMARY

● To draw a histogram with unequal class widths you need to calculate the frequency densities:

$$\text{Frequency density} = \frac{\text{frequency}}{\text{class width}}$$

The areas of the bars are then equal to the frequencies they represent.

● To read from a histogram you need to find the frequency:

$$\text{Frequency} = \text{frequency density} \times \text{class width}$$

QUICK TEST

1. The table opposite gives some information about the ages of some participants in a charity walk.

   On graph paper draw a histogram to represent this information.

Age (x) in years	Frequency
$0 < x \leqslant 15$	30
$15 < x \leqslant 25$	46
$25 < x \leqslant 40$	45
$40 < x \leqslant 60$	25

EXAM PRACTICE

1. The table and histogram give information about the distance (d km) travelled to work by some employees.

Distance (km)	Frequency	Frequency Density
$0 < d \leqslant 15$	12	
$15 < d \leqslant 25$		
$25 < d \leqslant 30$	36	
$30 < d \leqslant 45$		
$45 < d \leqslant 55$	20	

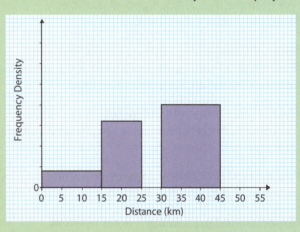

a. Use the information in the histogram to complete the table.

b. Use the table to complete the histogram.

2. The histogram shows the masses of onions in grams in a sack.

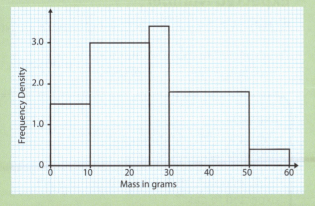

a. How many onions have a mass of between 10 and 25 grams?

b. How many onions were there in total?

Probability

Probability is the chance or likelihood that something will happen. All probabilities lie between 0 (impossible) and 1 (certain).

Probabilities must be written as a fraction, decimal or percentage.

Probability of a Single Event

$$P \text{ (event)} = \frac{\text{number of ways an event can happen}}{\text{total number of outcomes}}$$

● **Exhaustive events** account for all possible outcomes. For example, 1, 2, 3, 4, 5, 6 give all possible outcomes when a dice is thrown.

● **Mutually exclusive** events are events that cannot happen at the same time, e.g. a head and a tail on a fair coin cannot appear at the same time.

● Two or more events are **independent** when the outcome of the second event is not affected by the outcome of the first event.

● **Theoretical probability** analyses a situation mathematically.

● **Experimental probability** is determined by analysing the results of a number of trials or events. This is known as **relative frequency**. For example, a dice is thrown 55 times. A four comes up 13 times. The relative frequency is $\frac{13}{55}$.

Probability that an Event will NOT Happen

P (event will not happen) = 1 – P (event will happen)

Example

The probability that an alarm clock fails to go off is 0.21

What is the probability that the alarm clock will go off?

1 – 0.21 = **0.79**

Expected Number

Probability helps you predict the outcome of an event.

$$\text{The expected number of outcomes} = \text{number of trials} \times \text{probability}$$

Example

The probability of passing an exam in microbiology is 0.37

If 100 people take the exam, how many are expected to pass?

100 × 0.37 = **37 people**

Sample Space Diagrams

Sample space diagrams can be helpful when considering the outcomes of two events.

Example
Two spinners are spun and the scores added.

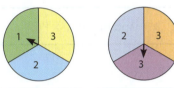

Represent the outcomes on a sample space diagram.

		Spinner 1		
		1	2	3
	2	3	4	5
Spinner 2	3	4	5	6
	3	4	5	6

QUESTIONS

QUICK TEST

1. The probability that Ahmed does his homework is 0.65

   What is the probability that Ahmed does not do his homework?

2. The probability of achieving a grade C in Maths is 0.49

   If 600 students sit this exam, how many would you expect to achieve a grade C?

3. Two fair dice are thrown and their scores are multiplied. By drawing a sample space diagram, what is:

   a. the probability of a score of 6?

   b. the probability of an even score?

EXAM PRACTICE

1. The probability that a factory manufactures a faulty component is 0.03

   Each day 1200 components are manufactured. Estimate the number of faulty components each day.

2. Imran plays a game of throwing a dart at a target. The table shows information about the probability of each possible score.

Score	0	1	2	3	4	5
Probability	0.04	x	$4x$	0.23	0.19	0.28

   Imran is 4 times more likely to score 2 points than to score 1 point. Work out the value of x.

Tree Diagrams

Tree diagrams are used to show the possible outcomes of two or more events. There are two rules you need to know first.

● **The OR rule**
If two events are mutually exclusive, the probability of A or B happening is found by adding the probabilities.

> P(A or B) = P(A) + P(B)

(These rules also work for more than two events.)

● **The AND rule**
If two events are independent, the probability of A and B happening together is found by multiplying the separate probabilities.

> P(A and B) = P(A) × P(B)

Example
A bag contains 3 red and 4 blue counters. A counter is taken from the bag at random, its colour is noted and then it is replaced in the bag. A second counter is then taken out of the bag. Draw a tree diagram to illustrate this information.

> Remember that the probabilities on the branches leaving each point on the tree add up to 1.

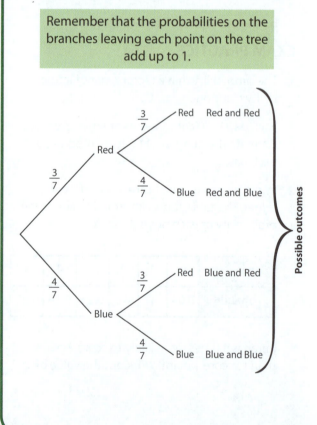

Work out the probability of:

a. picking two blues

> To find the probability of picking a blue AND a blue, multiply along the branches.

$$P(B) \times P(B)$$
$$= \frac{4}{7} \times \frac{4}{7} = \frac{\mathbf{16}}{\mathbf{49}}$$

b. picking one of either colour

> The probability of picking one of either colour is the probability of picking a blue and a red OR a red and a blue. You need to use the AND rule and the OR rule.

P(blue and red)

$$P(B) \times P(R)$$
$$= \frac{4}{7} \times \frac{3}{7} = \frac{12}{49}$$ The AND rule

P(red and blue)

$$P(R) \times P(B)$$
$$= \frac{3}{7} \times \frac{4}{7} = \frac{12}{49}$$ The AND rule

P(one of either colour)

$$= \frac{12}{49} + \frac{12}{49} = \frac{\mathbf{24}}{\mathbf{49}}$$ The OR rule

- If two events are mutually exclusive, the probability of A or B happening is found by adding the probabilities.

> P(A or B) = P(A) + P(B)

- If two events are independent, the probability of A and B happening is found by multiplying the probabilities.

> P(A and B) = P(A) × P(B)

- These rules also work for more than two events.
- Remember that the probabilities on the branches leaving each point on a tree diagram should add up to 1.

QUICK TEST

1. Sangeeta has a biased dice. The probability of getting a three is 0.4. She rolls the dice twice.

   a. Complete the tree diagram.

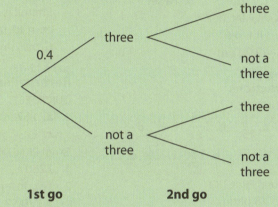

   1st go   **2nd go**

   b. Work out the probability that she gets:

      i. two threes

      ii. exactly one three

EXAM PRACTICE

1. A bag contains 3 red, 4 blue and 2 green beads. A bead is picked out of the bag at random and its colour noted. It is replaced in the bag. A second bead is picked out at random. Work out the probability that two different-coloured beads are chosen.

2. Mr Smith and Mrs Tate both go to the library every Wednesday. The probability that Mr Smith takes out a fiction book is 0.8, whilst the probability that Mrs Tate takes out a fiction book is 0.4. The events are independent.

   a. Calculate the probability that both Mr Smith and Mrs Tate take out a fiction book.

   b. Calculate the probability that one of each type of book is taken out.

Day 1

pages 4–5
Prime Factors, HCF and LCM
QUICK TEST
1. **a.** $50 = 2 \times 5^2$
   b. $360 = 2^3 \times 3^2 \times 5$
   c. $16 = 2^4$
2. **a.** False   **b.** True
   c. True   **d.** False
EXAM PRACTICE
1. HCF = 6
2. 12.20 pm

pages 6–7
Fractions
QUICK TEST
1. **a.** $\frac{13}{15}$ **b.** $2\frac{11}{21}$ **c.** $\frac{10}{63}$ **d.** $\frac{81}{242}$
2. **a.** $5\frac{7}{10}$ **b.** $1\frac{53}{90}$ **c.** $\frac{32}{75}$ **d.** 14

EXAM PRACTICE
1. 28 pages
2. $3\frac{5}{12}$ hrs = 3 hrs 25 minutes

pages 8–9
Percentages
QUICK TEST
1. **a.** 6 kg   **b.** £600
   c. £3   **d.** 252 g
2. **a.** 32%   **b.** 23%
   c. 75%   **d.** 84%
3. £180
4. 80%
EXAM PRACTICE
1. Best TV Shop £444.80, Drymons £445.50, Mark's Electricals £462. Since the television is the cheapest in Best TV Shop, this is where Jonathan should purchase the television.
2. £9410

pages 10–11
Repeated Percentage Change
QUICK TEST
1. 51.7% (3sf)
2. £121 856
EXAM PRACTICE
1. Savvy Saver – interest earned is £150, Money Grows – interest earned is £151.88
   Shamil is not correct – he would earn £1.88 more with the Money Grows investment.

pages 12–13
Reverse Percentage Problems
QUICK TEST
1. **a.** £57.50   **b.** £127
   c. £237.50   **d.** £437.50
EXAM PRACTICE
1. £250
2. Yes, Joseph is correct since $\frac{60}{0.85}$ = £70.59

pages 14–15
Ratio and Proportion
QUICK TEST
1. £20 : £40 : £100
2. £35.28
3. 3 days
EXAM PRACTICE
1. There are many ways to work this out – this is one possible way.
   Work out the cost of 25 ml for each tube of toothpaste.
   50 ml = £1.24: 25 ml = 62p
   75 ml = £1.96: 25 ml = 65.3̇p
   100 ml = £2.42: 25 ml = 60.5p
   The 100 ml tube of toothpaste is the better value for money.
2. Cheaper in America by £9.72 (or by $14.48)

Day 2

pages 16–17
Rounding and Estimating
QUICK TEST
1. **d.** 3700
2. **a.** True **b.** False **c.** True **d.** False
EXAM PRACTICE
1. $\dfrac{300 \times 3}{0.05} = \dfrac{900}{0.05} = \dfrac{90\,000}{5} = 18\,000$
2. 27.5 grams

pages 18–19
Indices
QUICK TEST
1. **a.** 6^8 **b.** 12^{13} **c.** 5^6 **d.** 4^2
2. **a.** $6b^{10}$ **b.** $2b^{-16}$ **c.** $9b^8$
   d. $\dfrac{1}{25x^4y^6}$ or $\dfrac{1}{25}x^{-4}y^{-6}$
EXAM PRACTICE
1. **a.** 1   **b.** $\dfrac{1}{49}$ **c.** 4   **d.** $\dfrac{1}{9}$
2. **a.** $x^{-4} = \dfrac{1}{x^4}$ **b.** $6x^3$

pages 20–21
Standard Index Form
QUICK TEST
1. **a.** 6.4×10^4 **b.** 4.6×10^{-4}
2. **a.** 1.2×10^{11} **b.** 2×10^{-1}
3. **a.** 1.4375×10^{18} **b.** 5.48×10^{19}
EXAM PRACTICE
1. **a.** 4×10^7 **b.** 0.00006
2. 1.4×10^{-6} g

pages 22–23
Formulae and Expressions 1
QUICK TEST
1. **a.** $4a$ **b.** $8a + b$
   c. $8a - 7b$ **d.** $15xy$
   e. $4a^2 - 8b^2$ **f.** $2xy + 2xy^2$
EXAM PRACTICE
1. **a.** $4bc - 2ab$
   b. d^4
   c. $15mn$
2. $T = 7x + 0.98y$
3. $C = 2x + 4y$

pages 24–25
Formulae and Expressions 2
QUICK TEST
1. **a.** $-\dfrac{31}{5} = -6\dfrac{1}{5}$ **b.** 4.36 or $\dfrac{109}{25}$

c. 9

2. $u = \pm\sqrt{v^2 - 2as}$

EXAM PRACTICE
1. $p = \dfrac{5a - 3b}{3}$

2. $b = \dfrac{m}{h^2}$ $m = 84.5$ $h = 1.79\,\text{m}$

$b = \dfrac{84.5}{1.79^2}$

$b = 26.37$

Yes, Dan would be classed as overweight.

pages 26–27
Brackets and Factorisation
QUICK TEST
1. **a.** $x^2 + x - 6$ **b.** $4x^2 - 12x$
   c. $x^2 - 6x + 9$
2. **a.** $6x(2y - x)$ **b.** $3ab(a + 2b)$
   c. $(x + 2)(x + 2) = (x + 2)^2$
   d. $(x + 1)(x - 5)$ **e.** $(x + 10)(x - 10)$

EXAM PRACTICE
1. **a.** $3t^2 - 4t$ **b.** $6x + 4$
2. **a.** $y(y + 1)$ **b.** $5pq(p - 2q)$
   c. $(a + b)(a + b + 4)$
   d. $(x - 2)(x - 3)$

pages 28–29
Equations 1
QUICK TEST
1. $x = 8$
2. $x = -5$
3. $x = -\dfrac{1}{2}$
4. $x = -3.25$
5. $x = -\dfrac{1}{2}$
6. $x = 17$

EXAM PRACTICE
1. **a.** $x = 2.4$ **b.** $x = -2.5$ **c.** $y = 2$
2. **a.** $x = -\dfrac{1}{5}$ **b.** $x = -4\dfrac{1}{3}$

Day 3
pages 30–31
Equations 2
QUICK TEST
1. $x = 5.5\,\text{cm}$, shortest length:
   $2x - 5 = 6\,\text{cm}$

2. **a.** $x = 0, x = 7$
   b. $x = -5, x = -3$
   c. $x = 3, x = 2$

EXAM PRACTICE
1. $x = 54$ smallest angle = 64°
2. $x = 2.7$

x	$x^3 + 4x^2 = 49$	Comment
2	$2^3 + 4 \times 2^2 = 24$	too small
3	$3^3 + 4 \times 3^2 = 63$	too big
2.5	$2.5^3 + 4 \times 2.5^2 = 40.625$	too small
2.7	$2.7^3 + 4 \times 2.7^2 = 48.843$	too small
2.75	$2.75^3 + 4 \times 2.75^2 = 51.046\ldots$	too big
2.74	$2.74^3 + 4 \times 2.74^2 = 50.601\ldots$	too big

pages 32–33
Simultaneous Linear Equations
QUICK TEST
1. $b = -4.5$ $a = 4$
2. $x = 2$ $y = 4$

EXAM PRACTICE
1. $a = 3$ $b = -2$
2. Hat £2   Balloon £3

pages 34–35
Sequences
QUICK TEST
1. **a.** $4n + 1$ **b.** $2 - n$ **c.** $2n$
   d. $3n + 2$ **e.** $5n - 1$

EXAM PRACTICE
1. **a.** $2n + 3$
   b. $2n^2 + 1 = 101$
      $2n^2 = 101 - 1$
      $2n^2 = 100$
      $n^2 = 50$
   Chloe is not correct. Since 50 is not a square number, 101 is not in the sequence.

pages 36–37
Inequalities
QUICK TEST
1. **a.** $x < 2.2$ **b.** $\dfrac{4}{3} \leqslant x < 3$
   c. $x > -\dfrac{9}{5}$

EXAM PRACTICE
1. **a.** $-2, -1, 0, 1, 2, 3, 4$ **b.** $x < 2$
2.

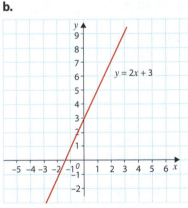

pages 38–39
Straight-line Graphs
QUICK TEST
1. **a.**

x	-2	-1	0	1	2	3
y	-1	1	3	5	7	9

b.

EXAM PRACTICE
1. **a.** $y = -2x + 3$ **b.** $y = \left(\dfrac{1}{2}, 1\right)$

pages 40–41
Curved Graphs
QUICK TEST
1. Graph A: $y = \dfrac{3}{x}$

   Graph B: $y = 4x + 2$

   Graph C: $y = x^3 - 5$

   Graph D: $y = 2 - x^2$

1. a.

x	−3	−2	−1	0	1	2	3
y	−28	−9	−2	−1	0	7	26

b.

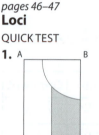

$y = x^3 - 1$

$y = 15$

c. $x = 2.5$

Day 4

pages 42–43
Distance–Time Graphs
QUICK TEST
1. 48 km/h
2. 3 mph

EXAM PRACTICE
1. **a.** False
   b. True
   c. True
   d. True
   e. True

pages 44–45
Constructions
QUICK TEST
1. Construct an equilateral triangle first. Bisect the 60° angle.

2. Perpendicular bisector of an 8 cm line should be drawn.

EXAM PRACTICE

1.

a.

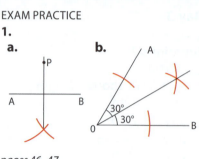

b.

A

30°
30°
0 ———— B

pages 46–47
Loci
QUICK TEST
1.

A ———— B

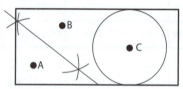

D ———— C

EXAM PRACTICE
1. a. b.

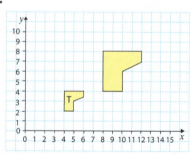

pages 48–49
Angles
QUICK TEST
1. **a.** $a = 55°$
   b. $a = 74°, b = 32°$
   c. $a = 41°, b = 41°, c = 41°, d = 139°$
2. 30°

EXAM PRACTICE
1. For BE and CF to be parallel, the angles between the parallel lines are supplementary and must add up to 180°.
   53° + 127° = 180°
2. $y = 149°$

pages 50–51
Bearings
QUICK TEST
1. **a.** 072° **b.** 208° **c.** 290° **d.** 139°

EXAM PRACTICE
1. **a.** 098° **b.** 187°
2. 323°

pages 52–53
Translations and Reflections
QUICK TEST
1. **a.** Reflection in $y = 0$ (x-axis)
   b. Reflection in $x = 0$ (y-axis)
   c. Translation of $\begin{pmatrix} -6 \\ 0 \end{pmatrix}$
   d. Translation of $\begin{pmatrix} 5 \\ -1 \end{pmatrix}$

EXAM PRACTICE
1. a. b.

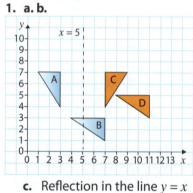

c. Reflection in the line $y = x$

Day 5

pages 54–55
Rotation and Enlargement
QUICK TEST
1.

2.

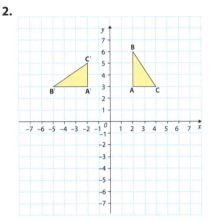

EXAM PRACTICE

1.

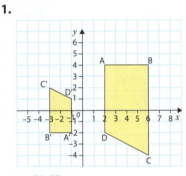

pages 56–57
Similarity
QUICK TEST
1. **a.** 13.8 cm (3sf)
   b. 4.5 cm (3sf)

EXAM PRACTICE
1. No, Lucy is not correct. The ratios of the corresponding lengths are not the same.

   $\frac{7}{3.5} = 2$ is not the same ratio as $\frac{3.7}{1.3}$

   $= 2.84…$
2. 20.7 cm (3sf)

pages 58–59
Circle Theorems
QUICK TEST
1. **a.** 62° **b.** 109° **c.** 53° **d.** 50°
   e. 126°

EXAM PRACTICE
1. **a.** 19° A tangent and radius meet at 90°.

b. 71° Angle in a semicircle is 90°.
   $E\hat{F}G = 90°$. Hence $G\hat{E}F = 71°$.

pages 60–61
Pythagoras' Theorem
QUICK TEST
1. **a.** 17.46 cm (2dp) **b.** 9.38 cm (2dp)

EXAM PRACTICE
1. Since $26^2 = 24^2 + 10^2$
   676 = 576 + 100 and this obeys Pythagoras' Theorem, then the triangle must be right-angled.
2. £15.66
3. $\sqrt{149} = 12.2$

pages 62–63
Trigonometry
QUICK TEST
1. **a.** 5.79 cm **b.** 8.40 cm
2. **a.** 38.7° **b.** 43.0°

EXAM PRACTICE
1. **a.** 57.4° (3sf)
   b. $\sin 48° = \frac{19.7}{PA}$ $PA = \frac{19.7}{\sin 48°}$

   $PA = 26.5$ m (3sf)

pages 64–65
Trigonometry and Pythagoras Problems
QUICK TEST
1. 1.95 m² (3sf)

EXAM PRACTICE
1. 3.06 m (2dp)
2. **a.** 12.7° (1dp)
   b. 4.1 m (1dp)
3. Bearing 218° (3sf)
   Distance 12.4 km (3sf)

Day 6

pages 66–67
Measurement
QUICK TEST
1. 3.2 kg
2. 8.8 lbs
3. 10.5 pints
4. 3 hrs 26 minutes

EXAM PRACTICE
1. 80 kg
2. 5 miles ≈ 8 km
   Josie: 70 mph = $70 \times \frac{8}{5}$

   $= 112$ km/h
   Jack: speed = 120 km/h
   Hence, Jack is travelling faster.

pages 68–69
Areas of Plane Shapes 1
QUICK TEST
1. **a.** 38 cm²   **b.** 33.58 cm²
   c. 35.705 cm²

EXAM PRACTICE
1. £2325

pages 70–71
Areas of Plane Shapes 2
QUICK TEST
1. **a.** Circumference = 31.42 cm (2dp)
   Area = 78.55 cm² (2dp)
   b. Circumference = 47.13 cm (2dp)
   Area = 176.74 cm² (2dp)
   c. Circumference = 43.99 cm (2dp)
   Area = 153.96 cm² (2dp)
2. Perimeter = 41.13 cm (2dp)
   Area = 100.53 cm² (2dp)

EXAM PRACTICE
1. 35 sacks of grass seed.

pages 72–73
Volumes of Prisms
QUICK TEST
1. **a.** 96 cm³ (1dp)
   b. 388.8 cm³ (1dp)
   c. 1292.3 cm³ (1dp)

EXAM PRACTICE
1. 8.32 cm (3sf)

pages 74–75
Pie Charts
QUICK TEST
1.

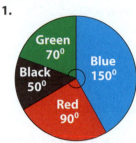

EXAM PRACTICE
1. Town D should have the retirement home since it has a much larger number of people aged 40 or over: 1100 people compared to 475 people in town C. Town C should have the school since it has 200 0–19 year-old people compared to 150 people in town D aged 0–19.

pages 76–77
Scatter Diagrams and Correlation
QUICK TEST
1. True   **2.** False   **3.** True
4. True   **5.** True
EXAM PRACTICE
1. **a.** Line of best fit should be as close as possible to all points and in the direction of the data.
   b. approx. 2 years old
   c. approx. £3000

Day 7
pages 78–79
Averages 1
QUICK TEST
1. **a.** Mean = 4.75
      Range = 7
   b. Mean = 9.5
      Range = 18
2. **a.** Median = 2.5
      Mode = 2
   b. Median = 6
      Mode = 6

EXAM PRACTICE
1. **a.** 6.93
   b. 7
   c. 7
   d. 5

pages 80–81
Averages 2
QUICK TEST
1. 150.9
2. a.

```
0 | 7 9
1 | 8
2 | 5 5 5 6 7 7 8
3 | 1 1 1 3 6 6 7 9
4 | 0 2 2 7 9
5 | 0
```
   Key 4|2 = 42
   b. Median = 31
EXAM PRACTICE
1. 5.52 hours

pages 82–83
Cumulative Frequency Graphs
QUICK TEST
1.

	Time in Minutes
Median time	17
Lower quartile	12
Interquartile range	11
Longest time	29

EXAM PRACTICE
1. **a.** 9B
   b. Class 9A. The median is higher than 9B; 50% of students in 9A scored over 65 marks. The top 50% of students in class 9B scored over 55 marks.

pages 84–85
Histograms
QUICK TEST
1.

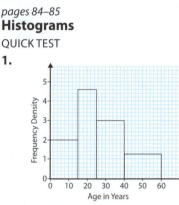

EXAM PRACTICE
1.a.

Distance (km)	Frequency	Frequency Density
$0 < d \leqslant 15$	12	0.8
$15 < d \leqslant 25$	32	3.2
$25 < d \leqslant 30$	36	7.2
$30 < d \leqslant 45$	60	4.0
$45 < d \leqslant 55$	20	2.0

b.

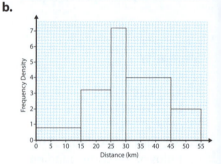

2. **a.** 45 onions
   b. 117 onions

pages 86–87
Probability
QUICK TEST
1. 0.35
2. 294 students
3. a. $\frac{4}{36} = \frac{1}{9}$ **b.** $\frac{27}{36} = \frac{3}{4}$

EXAM PRACTICE
1. 36
2. 0.052

pages 88–89
Tree Diagrams

QUICK TEST

1. a.

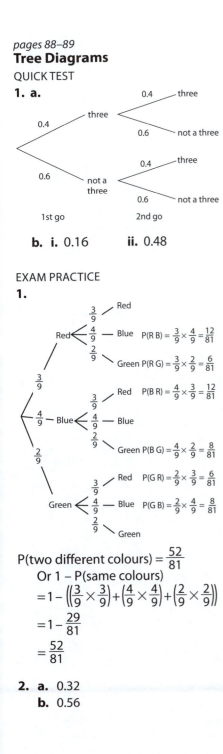

1st go   2nd go

b. i. 0.16   **ii.** 0.48

EXAM PRACTICE

1.

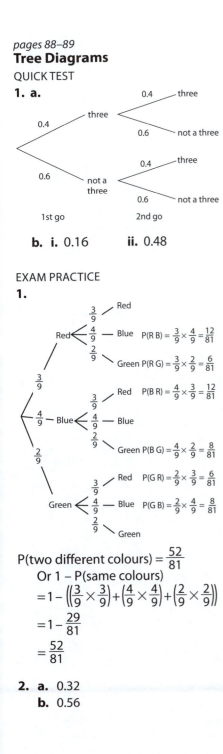

$P(\text{two different colours}) = \dfrac{52}{81}$

Or $1 - P(\text{same colours})$

$= 1 - \left(\left(\dfrac{3}{9} \times \dfrac{3}{9}\right) + \left(\dfrac{4}{9} \times \dfrac{4}{9}\right) + \left(\dfrac{2}{9} \times \dfrac{2}{9}\right)\right)$

$= 1 - \dfrac{29}{81}$

$= \dfrac{52}{81}$

2. a. 0.32

    b. 0.56

Notes